AF430918

Solved Problems in
Geotechnical Engineering

Solved Problems in Geotechnical Engineering

Ravi Kumar Sharma

Rachit Sharma

BSP **BS Publications**

An Imprint of **BSP Books Pvt. Ltd.**
4-4-309/316, Giriraj Lane,
Sultan Bazar, Hyderabad - 500 095

Solved Problems in Geotechnical Engineering
by Ravi Kumar Sharma and Rachit Sharma

© 2024, *by Publisher*

Disclaimer: The authors and the publishers have taken due care to provide the authentic, reliable and up to date information related to the subject. However, neither the authors nor the publisher shall be responsible for any liability for any damage caused as a result of use of this book. The respective user must check the accuracy from other sources too.

Published by:

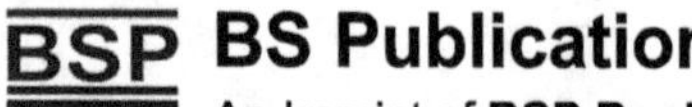 **BS Publications**

An Imprint of **BSP Books Pvt. Ltd.**
4-4-309/316, Giriraj Lane, Sultan Bazar,
Hyderabad - 500 095
Phone: 040 - 23445688
e-mail : info@bspbooks.net
www.bspbooks.net

ISBN: 978-93-95038-48-5 (Hardback)

Preface

This text book on Problems in Geotechnical Engineering has been written with a view to provide thorough understanding of problems in the subject to the readers. A large number of problems of varied types have been solved in different chapters of Geotechnical Engineering subject generally offered at the undergraduate level. Suitable diagrams and illustrations have been included in the appropriate problems. The numerical problems from the previous question papers of competitive examinations e.g., GATE, engineering services and administrative services have been solved and presented in the book. Sufficient number of unsolved problems have been given at the end of each chapter to enable the students prepare for different examinations. The book will provide comprehensive knowledge to the students in the subject. Although utmost care has been taken to avoid any errors, however, comments and criticism are welcomed from the readers.

The authors are especially thankful to different individuals and groups for providing all support and help in this endeavour.

- Authors

Contents

CHAPTER 1

Properties of Soil

Question 1.1: The total unit weight of the glacial outwash soil is 16 kN/m^3. The specific gravity of soil particles of the soil is 2.67. The water content of the soil is 17%. Calculate

(a) Dry unit weight

(b) Porosity

(c) Void ratio

(d) Degree of saturation

Solution: Total unit weight $\gamma = 16$ kN/m^3

Specific gravity $G = 2.67$

Water content $w = 17\%$

Dry unit weight $\gamma_d = \gamma/(1+w) = 16/(1+0.17) = 13.675$ kN/m^3

Void ratio $e = (G\gamma_w/\gamma_d) - 1 = (2.67 \times 9.81/13.675) - 1 = 0.9153$

Porosity $n = e/(1+e) = 0.9153/(1+0.9153) = 47.79\%$

Degree of saturation $S = wG/e = 0.17 \times 2.67/.9153 = 49.59\%$

Question 1.2: The porosity (n) and the degree of saturation (S) of a soil sample are 0.7 and 40%, respectively. Determine the volume (in m^3) of air in 100 m^3 volume of the soil.

Solution: Porosity $= n = V_v/V = 0.7$

Volume of voids $V_v = 0.7 \times 100 = 70$ m^3

$V_w/V_v =$ degree of saturation $S = 0.4$

Volume of water $V_w = 70 \times 0.4 = 28$ m^3

Volume of air $V_a = V_v - V_w = 70 - 28 = 42$ m^3

Volume of air $= 42$ m^3

Question 1.3: A soil sample has specific gravity of 2.60 and void ratio of 0.78. Determine the water content in percentage required to fully saturate the soil at that void ratio.

Solution: Degree of saturation $S = 100\%$

$$\text{Water content } w = Se/G$$
$$= 100 \times 0.78/2.60$$
$$= 30\%$$

Question 1.4: A soil has mass unit weight γ, water content w (expressed as ratio). The specific gravity of soil solids G, unit weight of water $= \gamma_w$, then show that the degree of saturation of soil S is given by: $S = w/ [(\gamma_w/\gamma)(1+w) - 1/G]$

Solution: $\gamma = G(1+w)\gamma_w /(1+e)$

$$1+e = G(1+w)\gamma_w/\gamma$$

and $\quad Gw = Se$

$$Gw/S = [G(1+w)\gamma_w/\gamma] - 1$$

Dividing both sides by G, we get

$$w/S = [(1+w)\gamma_w/\gamma] - 1/G$$
$$S = w/ [(\gamma_w/\gamma)(1+w) - 1/G]$$

Question 1.5: A sample of saturated sand has a dry unit weight 18 kN/m^3 and a specific gravity of 2.7. If the unit weight of water is 10 kN/m^3, determine the void ratio of soil sample.

Solution: Dry unit weight $\gamma_d = G\gamma_w/(1+e)$

$$\text{Void ratio } e = (G\gamma_w/ \gamma_d) - 1$$
$$= 2.7 \times 10/18$$
$$= 0.5$$

Question 1.6: The natural void ratio of sand sample is 0.6 and its density index is 0.6. If the void ratio in the loosest state is 0.9, then determine the void ratio in its densest state.

Solution: Density index $= (e_{max} - e)/(e_{max} - e_{min}) = 0.6$

Void ratio in loosest state,

$$e_{max} = 0.9$$

Void ratio in natural state,

$$e = 0.6$$
$$0.9 - e_{min} = (0.9 - 0.6)/ 0.6$$
$$e_{min} = 0.9 - 0.5$$
$$= 0.4$$

Question 1.7: What will be the relative density of saturated sand deposit having moisture content of 25%, maximum and minimum void ratio 0.95 and 0.45 respectively and specific gravity of particles 2.6?

Solution: At saturated moisture content void ratio is

$$e = Gs/S$$

$$= 2.6 \times 25/100 = 0.65$$

Relative density

$$D_r = (e_{max} - e)/(e_{max} - e_{min})$$

$$= (0.95 - 0.65)/(0.95 - 0.45) = 0.6$$

$$D_r = 60\%.$$

Question 1.8: A given cohesionless soil has $e_{max} = 0.85$ and $e_{min.} = 0.50$. In the field, the soil is compacted to a mass density of 1800 kg/m^3 at a water content of 8%. Take the mass density of water as 1000 kg/m^3 and G_s as 2.7; determine the relative density (in %) of the soil.

Solution: Given, soil with the field properties

Bulk density $(\rho) = 1800$ kg/m^3

Water content, w = 8% = 0.08

So dry density $\rho_d = \rho/(1+w) = 1800/(1+0.08)$

$$= 1666.67 \text{ kg/m}^3$$

So void ratio 'e' at the field condition can be determined by relation

$$\rho_d = G\rho_w/(1+e) \Rightarrow e = \{G\rho_w/(\rho_d)\} - 1$$

$$= \{(2.7 \times 1000)/(1666.67)\} - 1$$

$$e = 0.62$$

So $\quad e_{max} = 0.85$, $e_{min} = 0.50$

Relative density $I_d = \{(e_{max} - e)/(e_{max} - e_{min})\} \times 100$

$$= \{(0.85-0.62)/(0.85-0.50)\} \times 100$$

$$I_d = 65.71\%$$

Question 1.9: What are the respective values of void ratio, porosity ratio and saturated unit weight (in kN/m^3) for a soil sample which has saturation moisture content of 20% and specific of grains 2.6. Take unit weight of water 10kN/m^3.

Solution:

Given, water content w = 20%,

Specific gravity G = 2.6,

Degree of saturation S = 100%

Void ratio   e = wG/S = (20×2.6)/100 = 0.52

Porosity   n = e/(1+e) = 0.34

Saturated unit weight γ_{sat} = [(G + e)/ (1 + e)] γ_w

$$= [(2.6 + 0.52)/(1 + 0.52)] \times 10$$

$$= 20.53 \text{ kN/m}^3$$

If unit weight of water is taken as 9.81 kN/m³

Then γ_{sat} = [(2.6 + 0.52)/(1 + 0.52)] ×9.81

$$= 20.14 \text{ kN/m}^3$$

Question 1.10: What is the dry unit weight of clay soil when the void ratio of sample thereof is 0.50, the degree of saturation is 70% and specific gravity of soil grains is 2.7? Take the value of γ_w to be 9.81 kN/m³.

Solution: Bulk unit weight

$$\gamma_b = (G + Se) \gamma_w/ (1+ e)$$

$$= [(2.7 + 0.7 \times 0.5) \times 9.81]/(1+ 0.5) \text{ kN/m}^3$$

$$= 19.947 \text{ kN/m}^3$$

$$Se = wG$$

Water content w = Se/G

$$w = 0.7 \times 0.5/2.7 = 0.1296 = 12.96\%$$

Dry density $\gamma_d = \gamma_b/ (1 + w) = 19.947/ (1 + 0.1296)$

$$= 17.658 \text{ kN/m}^3.$$

Question 1.11: An earth embankment is to be constructed with compacted cohesionless soil. The volume of the embankment is 5000 m³ and the target dry unit weight is 16.2 kN/m³. Three nearly sites (see figure below) have been identified from where the required soil can be transported to the constructed site. The void ratios (e) of different sites are shown in the figure below. Assume the specific gravity of soil to be 2.7 for all three sites. If the cost of transportation per km is twice the cost of excavation per m³ of borrow pits, which site would you choose as the most economical solution? Take unit weight of water = 10 kN/m³.

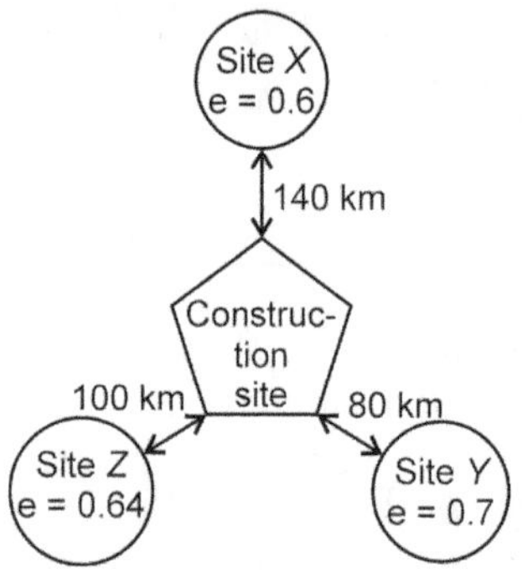

Figure 1.1

Solution:

Volume of solids at embankment,

Volume of solids $V_s = W_s/ G_s\gamma_\omega$

$$V_s = \frac{(16.2 \times 5000)}{(2.7 \times 10)}\, m^3$$

$V_s = 3000 \ m^3$

Volume of solids will remain constant

Also, total volume $V = V_v + V_s = eV_s + V_s$

$$= V_s\,(1 + e)$$

$V = 3000(1 + e)$

For site X; $V_x = 3000 \times 1.6 = 4800 \ m^3$

For site Y; $V_y = 3000 \times 1.7 = 5100 \ m^3$

For site Z; $V_z = 3000 \times 1.64 = 4920 \ m^3$

Let cost of excavation be α/m^3,

$P_x = 4800\ \alpha + 140 \times \alpha \times 2$

$\quad = 5080\ \alpha$

$P_y = 5100\ \alpha + 80 \times \alpha \times 2$

$\quad = 5260\ \alpha$

$P_z = 4920\ \alpha + 100 \times \alpha \times 2$

$\quad = 5120\ \alpha$

Here P_x is least.

Therefore, total cost for site X is the least.

Question 1.12: A fine-grained soil has 60% (by weight) silt content. The soil behaves as semi-solid when water content is between 15% and 28%. The soil behaves fluid-like when the water content is more than 40%. Determine the activity of the soil.

Solution: Using index properties of soil

Shrinkage limit $w_s = 15\%$

Plastic limit $w_p = 28\%$

Liquid limit $w_l = 40\%$

Plasticity index $I_p = W_l - W_p = 12\%$

Activity of soil, $A = I_p/c = 12/ (100\text{-}60)$

$$= 0.3.$$

Question 1.13: A soil sample has shrinkage limit of 6% and the specific gravity of soil grains is 2.6. Determine the porosity of soil at shrinkage limit.

Solution: Shrinkage limit $w_s = 6\%$;

Specific gravity $G = 2.6$

At shrinkage limit, soil is fully saturated, hence $S = 1$

$$Se = wG$$

$$1 \times e = 0.06 \times 2.6$$

Void ratio $e = 0.156$

Porosity $n = e/(1 + e) = 0.156 /(1 + 0.156)$

$$= 0.1349$$

$$= 13.49\%$$

Question 1.14: The plastic limit and liquid limit of soil are 30% and 42% respectively. The percentage volume change from liquid limit to dry state is 35% of dry volume. Similarly, the percentage volume change from plastic limit to dry state is 22% of the dry volume. Determine the shrinkage ratio.

Solution: Given data:

Plastic limit $w_p = 30\%$

Liquid limit $w_l = 42\%$

$$V_1 - V_d = 0.35V_d$$

$$V_p - V_d = 0.22V_d$$

$$V_1 = 1.35V_d$$

$$V_p = 1.22V_d$$

$$SR = [(V_1 - V_2)/V_d]/(w_1 - w_2) \times 100$$

$$= [(V_1 - V_p)/ V_d]/(w_1 - w_p) \times 100$$

$$= [(1.35V_d - 1.22V_d)/V_d]/(42 - 30)$$

$$SR = (0.13/12) \times 100 = 1.083 \simeq 1.1$$

Question 1.15: A sample of saturated soil has a water content of 40%. The specific gravity G of soil particles is 2.7. Determine

1. void ratio
2. porosity
3. saturated and dry unit weights.

Solution: Given data:

Water content $w = 40\% = 0.40$,

Specific gravity $G = 2.7$

Void ratio, e

$$e = w_{sat}\, G$$
$$= 0.40 \times 2.7 = 1.08$$

Porosity, n

$$n = e/(1 + e)$$
$$= 1.081 / 2.081 = 0.519$$

Dry unit weight,

$$\gamma_d = (1 - n)\, G\, \gamma_w$$
$$= (1 - 0.519) \times 2.7 \times 9.81$$
$$= 12.74 \text{ kN/m}^3$$

Saturated unit weight, $\gamma_{sat} = \gamma_d + n\, \gamma_w$

$$= 12.74 + 9.81 \times 0.519 = 17.831 \text{ kN/m}^3.$$

Question 1.16: A certain sample of saturated soil with watch glass weighs 65.42g. On oven drying, the sample with watch glass is 60.85 g. The weight of watch glass is 35.45 g. The specific gravity of solid is 2.8. Determine

1. void ratio
2. water contents
3. porosity
4. unit weights

Solution: Given data:

Mass of wet soil, $M = 65.42 - 35.45 = 29.97$ g

Mass of dry soil, $M_d = 60.85 - 35.45 = 25.40$ g

Specific gravity, $G = 2.8$

Water content $w = [(M/M_d) - 1] \times 100\%$
$$= [(29.97/25.40) - 1] \times 100\%$$
$$= 17.99\%$$

Void ratio, $e = w_{sat}\, G$
$$= 0.1799 \times 2.8$$
$$= 0.504$$

Porosity, $n = e/(1 + e)$
$$= 0.504/1.504$$
$$= 0.335$$

Bulk unit weight, $\gamma = \gamma_w(G + Se)/(1 + e)$

$$= 9.81 \times (2.8+0.504\times1)/(1 +0.504)$$

$$= 21.551 \text{ kN/m}^3.$$

Dry unit weight, $\gamma_d = G\gamma_w(1 - n)$

$$= 9.81\times2.8\,(1 - 0.335) = 18.266 \text{ kN/m}^3.$$

Unit weight of solids, $\gamma_s = G\gamma_w$

$$= 9.81\times2.8 = 27.468 \text{ kN/m}^3.$$

Saturated unit weight, $\gamma_{sat} = \gamma_d + n\gamma_w$

$$= 18.266+9.81\times0.335 = 21.552 \text{ kN/m}^3.$$

Submerged unit weight, $\gamma' = \gamma_{sat} - \gamma_d$

$$= 21.552 - 9.81 = 11.742 \text{ kN/m}^3.$$

 **Questions**

Question 1: Soil is to be excavated from a borrow pit which has density of 1.75 g/cm^3 and water content of 12%. The specific gravity of soil particles is 2.7. The soil is compacted so that water content is 18% and dry density is 1.65 g/cm^3. For 1000 m^3 of soil in fill, estimate:

1. Quantity of soil to be excavated from the pit in m^3.
2. Amount of water to be added.
3. Void ratio of soil in borrow pit and fill.

(ESE: 1995)

Question 2: The values of liquid limit, plastic limit and shrinkage limit of a soil were reported as $w_l = 60\%$, $w_p = 30\%$, $w_s = 20\%$. If a sample of this soil at liquid limit has a volume of 40 cm^3 and its volume measured at shrinkage limit was 23.5 cm^3, determine the specific gravity of solids. What is the shrinkage ratio and volumetric shrinkage?

(ESE: 1996)

Question 3: The void ratio and specific gravity of a sample of clay are 0.73 and 2.7 respectively. If the voids are 92% saturated, find the bulk density, the dry density and water content. What would be the water content for complete saturation, the void ratio remaining the same?

(ESE: 1999)

Question 4: A sample of sand above water table was found to have a natural moisture content of 15% and a unit weight 18.84 kN/m^3. Laboratory tests on a drained sample indicated values of 0.5 and 0.85 for minimum and maximum void ratio respectively for the densest and the loosest states. Determine the degree of saturation and relative density. Assume $G = 2.65$.

(ESE: 2000)

Question 5: A solid sample has a porosity of 40%. The specific gravity of solid is 2.7. Calculate the (a) void ratio (b) dry density (c) unit weight if soil is 50% saturated and (d) unit weight if the solid is completely saturated.

(ESE: 2012)

Question 6: An oven-dry soil sample of volume 225 cm^3 weighs 390 g. If the specific gravity of soil is 2.72, determine the void ratio and shrinkage limit. What will be the water content which will fully saturate the sample and cause an increase in volume equal to 8% of the original dry volume?

(ESE: 2011)

Question 7: A solid sample has a porosity of 40%. The specific gravity of solid is 2.7. Calculate (a) void ratio (b) dry density (c) unit weight if the soil is 50% saturated and (d) unit weight if the solid is completely saturated.

(ESE: 2012)

Question 8: The mass of saturated soil sample is 150 g and its mass when oven dried is 90 g; find the water content. Suppose that the sample used in triaxial test has a diameter 38 mm and height 76 mm, determine the void ratio. (ESE: 2013)

Question 9: A soil deposit has a void ratio of 0.9. Its void ratio is reduced to 0.6 by compaction. Determine the percentage reduction of volume due to the compaction.

(ESE: 2014)

Question 10: A sampler with a volume of 45 cm^3 is filled with a soil sample. When the soil is poured into a graduated cylinder, it displaces 25 cm^3 of water. What is the porosity and void ratio of the soil?

(ESE: 1998)

Question 11: The water content of a saturated soil and the specific gravity of soil solids were found to be 30% and 2.70, respectively. Assuming the unit weight of water to be 10 kN/m^3, determine the saturated unit weight and the void ratio of the soil.

Question 12: Soil has been compacted is an embankment at a bulk density of 2.15 Mg/m^3 and a water content of 12%. The value of specific gravity of soil solids is 2.65. The water table is well below the foundation level. Estimate the dry density, void ratio, degree of saturation and air content of the compacted soil.

Question 13: A soil has a void ratio of 0·70, degree of saturation 50% and Gs = 2·7. Find the water content, porosity, bulk density and dry density. By how much can the water content be increased without changing the volume.

Question 14: A proposed earthen dam will have a volume of 5000000 m^3 of compacted soil. The soil is to be taken from a borrow pit and will be compacted to a void ratio of 0·8. The void ratio of soil in the borrow pit is 1·15. Estimate the volume of soil that must be excavated from the borrow pit for the construction of the above dam.

Question 15: A soil deposit has a void ratio of 0·9. Its void ratio is reduced to 0·6 by compaction. Determine the percentage reduction of volume by this compaction.

Question 16: The mass of the saturated soil sample is 150 gm and its mass when oven dried is 90 gm, find the water content. Suppose that the sample, used for triaxial test, has a diameter of 38 mm and the height of 76 mm, find the void ratio.

CHAPTER 2

Index Properties

Question 2.1: The liquid limit and plastic limit of soil sample are 65% and 75% respectively. The percentage of soil fraction with grain size finer than 0.002 mm is 24. Determine the activity of soil sample:

Solution: Activity = Plasticity index / percentage of clay particle finer than 2 μm

$$= (65 - 29) / 24$$

$$= 36/24$$

$$= 1.5 > 1.25$$

Hence, soil is active.

Question 2.2: Given that the plasticity index (PI) of local soil = 15% and PI of sand = 0, for a desired PI of 6, determine the percentage of sand in the mix.

Solution: Let the percentage of sand is x. So, plasticity of mix is

$$x \times 0 + (1 - x) \times 15 = 6$$

$$x = 9/15$$

$$= 0.6 \text{ or } 60\%$$

Question 2.3: If a soil sample is of weight 0.18 kg having a volume of 10^{-4} m^3 and dry unit weight of 1600 kg/m^3 is mixed with 0.02 kg of water. Determine the water content in the sample.

Solution: Dry weight of sample = $1600 \times 10^{-4} = 0.16$ kg

Weight of water in soil before mixing additional quantity

$$= 0.18 - 0.16 = 0.02 \text{ kg}$$

After mixing water the total quantity of water

$$= 0.02 + 0.02$$

$$= 0.04 \text{ kg of water}$$

Thus, water content = $(0.04/0.16) \times 100 = 25\%$

Question 2.4: A soil has a liquid limit of 40% and plasticity index of 20%. Determine the plastic limit of the soil.

Solution: Given,

$$w_p = w_l - PI$$
$$= 40 - 20 = 20\%$$

Question 2.5: A sample has a void ratio 0.54 in dry state. The specific gravity of soil solid is 2.7. What is the shrinkage limit of soil?

Solution: At shrinkage limit the soil remains fully saturated,

$$w_s = (e_{min}/G) \times 100$$
$$= (0.54/2.7) \times 100$$
$$= 20\%$$

Question 2.6: A soil has liquid limit = 35, plastic limit = 20, shrinkage limit = 10 and natural moisture content = 25%. What will be the liquidity index, plasticity index and shrinkage index?

Solution: Plasticity index, $I_p = w_l - w_p$

$$= 35 - 20 = 15\%$$

Liquidity index, $I_l = (w_n - w_p)/I_P$

$$= (25 - 20)/15 = 0.33$$

Shrinkage index, $I_s = w_p - w_s = 20 - 10 = 10\%$

Question 2.7: Consider the following properties for clays X and Y:

S. No.	Properties	Clay X (%)	Clay Y (%)
1.	Liquid limit	42	56
2.	Plastic limit	20	34
3.	Natural water content	30	50

Which of the clay, X and Y, experiences large settlement under identical loads; is more plastic; and is softer in consistency?

Solution: Plasticity index (I_p) is the numerical difference between liquid limit and plastic limit. The soil with higher value of I_p will be more plastic.

$$(I_p)_x = 42 - 20 = 22$$
$$(I_p)_y = 56 - 34 = 22$$

But if plasticity index is same for both soils, then soil with larger difference in liquid limit and natural water content is more plastic.

Consistency index, $I_c = $ (Liquid limit – natural water content) $/ I_p$

$$(I_p)_x = (42 - 30) / 22 = 0.545$$
$$(I_p)_y = (56 - 50) / 22 = 0.273$$

Soil which has low value of consistency index i.e., clay Y is softer. Thus, the softer soil i.e., clay Y will undergo large settlements under identical loads.

Question 2.8: A specimen of clayey silt contains 70% silt size particles. Its liquid limit is 40% and plastic limit is 20%. In liquid limit test, at moisture content of 30%, required number of blows was 50. Determine its plasticity index, activity and consistency index.

Solution: Plasticity index, $(I_p) = w_l - w_p$

$$= 40 - 20 = 20\%$$

Activity of clay,

A_c = Plasticity index / percent finer than 2μm

   = Plasticity index / percent of clay particles in the soil

   $= 20/ (100 - 70) = 0.67$

Consistency index,

$$I_c = (w_l - w_n)/I_p = 40 - 30/ 20 = 0.50.$$

Question 2.9: A saturated specimen of clay was immersed in mercury and its displaced volume was 21.8 cm^3. The weight of sample was 32.2 g. After oven drying for 48 hours, the weight reduced to 20.2 g, while the volume came down to 11.6 cm^3. Determine the shrinkage limit of soil.

Solution: Mass of water in initial condition

$$= 32.2 - 20.2 = 12 \text{ g}$$

Loss of water from original to shrinkage limit

$$= (21.8 - 11.6) \times 1 = 10.2 \text{ g}$$

Shrinkage limit $= (32.2 - 20.2 - 10.2) / 20.2 \times 100 = 8.91\%$

Question 2.10: A sample of dry soil is coated with a thin layer of paraffin and has a mass of 460 g. it displaced 300 cm^3 of water when immersed in it. The paraffin is peeled off and its mass was found to be 9 g. If the specific gravity of soil solids and paraffin are 2.65 and 0.9 respectively, determine the voids ratio of soil.

Solution: Volume of wax = 9/G = 9/0.9 = 10 cm^3

   Weight of dry soil = weight of solids = w_s = 451 g

   Volume of soil = 300 − 10 = 290 cm^3

   Volume of solid = V_s = W_s/G = 451/2.65 = 170.18 cm^3

   Volume of air = volume of void

$$= V_v = 290 - 170.18$$

$$V_v = 119.81 \text{ cm}^3$$
$$e = V_v/V_s = 119.81/170.18 = 0.70$$
$$e = 0.70$$

Question 2.11: The plastic limit and liquid limit of soil are 30% and 42% respectively. The percentage volume change from liquid limit to dry state is 35% of the dry volume. Similarly, the percentage volume change from plastic limit to dry state is 22% of the dry volume. Determine the shrinkage ratio.

Solution: Given data,

$$w_p = 30\%$$
$$w_l = 42\%$$
$$V_1 - V_d = 0.35V_d$$
$$V_p - V_d = 0.22V_d$$
$$V_1 = 1.35V_d$$
$$V_p = 1.22V_d$$
$$SR = [(V_1 - V_d)/V_d] / (w_1 - w_2) \times 100$$
$$= [(V_1 - V_p)/V_d] / (w_1 - w_p) \times 100$$
$$= [(1.35\,V_d - 1.22\,V_d)/V_d] / (42 - 30) \times 100$$
$$SR = 0.13/12 \times 100 = 1.083 \approx 1.1$$

Question 2.12: Determine the specific gravity of soil from the following data:

Mass of pycnometer $M_0 = 356$ g.

Mass of pycnometer + dry soil $M_1 = 455$ g.

Mass of pycnometer + soil + water $M_2 = 1025$ g.

Mass of pycnometer + water $M_3 = 963$ g.

Solution:

Specific gravity of soil grains

$$G = \frac{M1 - M0}{(M1 - M0) - (M2 - M3)} = \frac{455 - 356}{(455 - 356) - (1025 - 963)}$$
$$= 2.676$$

Question 2.13: Determine the plastic limit of soil from the following data:

Weight of container $W_1 = 15.32$g.

Weight of container + wet soil $W_2 = 22.02$g.

Weight of container + oven dry soil $W_3 = 20.76$g.

Solution:

Water content $= \dfrac{W2 - W3}{W3 - W1} \times 100 = \dfrac{22.02 - 20.76}{20.76 - 15.32} \times 100 = 23.16\%.$

Question 2.14: Determine the shrinkage limit of soil from the following data:

Mass of shrinkage dish $M_1 = 25.4$g.

Mass of shrinkage dish + wet soil pat $M_2 = 66.2$g.

Mass of shrinkage dish + dry soil pat $M_3 = 56.0$ g.

Mass of evaporating dish $M_e = 88.5$ g

Mass of mercury filling shrinkage dish + mass of evaporating dish

$\qquad M_{me} = 360.9$ g

Mass of mercury displaced by dry soil pat + mass of evaporating dish

$\qquad M_{mde} = 308.9$ g.

Also determine the shrinkage ratio and volumetric shrinkage.

Solution:

Mass of wet soil pat $M = M_2 - M_1 = 40.8$ g

Mass of dry soil pat $M_d = M_3 - M_1 = 30.6$ g

Water content $w = \left[\dfrac{M - M_d}{M_d}\right] \times 100 = \left[\dfrac{40.8 - 30.6}{30.6}\right] \times 100 = 33.33\%$

Mass of mercury filling shrinkage dish

$\qquad M_m = M_{me} - M_e = 360.9 - 88.5 = 272.4$ g

Volume of wet soil pat $V = M_m / 13.6 = 272.4/13.6 = 20.03$ cm^3

Mass of mercury displaced by dry soil pat

$\qquad M_{md} = M_{mde} - M_e = 308.9 - 88.5 = 220.4$ g

Volume of dry soil pat $V_d = M_{md} / 13.6 = 220.4/13.6 = 16.21$ cm^3

Shrinkage limit $w_s = w - \left[\dfrac{V - V_d}{M_d}\right] \times 100 = w - \left[\dfrac{20.03 - 16.21}{30.60}\right] \times 100$

$\qquad\qquad = 33.33 - 12.48 = 20.85\%$

Shrinkage ratio $SR = M_d / V_d = 30.60/16.21 = 1.89$

Volumetric shrinkage $VS = (w - w_s)\, SR = (33.33 - 20.85) \times 1.89 = 23.59\%.$

Question 2.15: Determine the dry density of soil from the following data:

Volume of core cutter $V_c = 1000$ cm^3.

Weight of core cutter + wet soil $W_s = 2250$ g.

Weight of core cutter $W_c = 436$ g.

Weight of water content container with lid $W_1 = 12.70$ g.

Weight of water content container with lid and wet soil $W_2 = 50.20$ g.

Weight of water content container with lid and dry soil $W_3 = 45.10$ g.

Solution:

Weight of wet soil $= W_s - W_c$

$$= 2250 - 436 = 1814 \text{ g.}$$

Bulk density, $\gamma_t = \dfrac{W_s - W_c}{V_c}$

$$= 1814/1000 = 1.814 \text{ g/cm}^3.$$

Water content $w = \dfrac{W_2 - W_3}{W_3 - W_1} \times 100$

$$= \dfrac{50.20 - 45.10}{45.10 - 12.70} \times 100 = 15.74\%.$$

Dry density $\gamma_d = \dfrac{\gamma_t}{1+w} \times 100 = 1.814/(1 + 15.74) = 1.567 \text{ g/cm}^3.$

Question 2.16: Results obtained from the sieve analysis of a sand sample are given below. Determine effective size, coefficient of uniformity and coefficient of curvature of sand by suitably plotting the results.

S. No.	Mesh opening (mm)	Weight retained (g)
1.	4.75	0
2.	2.36	0
3.	1.18	55.0
4.	0.600	85.0
5.	0.425	32.0
6.	0.300	109.5
7.	0.150	129.0
8.	0.750	19.0
9.	Pan	20.5

Solution: The results are analysed as given below:

Size of sieve opening D (mm)	Mass of soil retained (g)	Percentage retained (%)	Cumulative percentage retained (%)	Percentage finer N (%)
4.75	0	0	0	100
2.36	0	0	0	100
1.18	55.0	12.22	12.22	87.78
0.600	85.0	18.88	31.10	68.90

Contd….

Size of sieve opening D (mm)	Mass of soil retained (g)	Percentage retained (%)	Cumulative percentage retained (%)	Percentage finer N (%)
0.425	32.0	7.11	38.21	61.79
0.300	109.5	24.33	62.54	37.46
0.150	129.0	28.66	91.20	8.80
0.750	19.0	4.22	95.42	4.58
Pan	20.5	4.58	100	0
Total	= 450. 0	100.00		

Sample No.	D_{60} (mm)	D_{30} (mm)	D_{10} (mm)	C_U	C_C	Cobble (%)	Gravel (%)	Sand (%)	Silt or clay (%)
1	0.41	0.25	0.17	2.41	0.90	-	-	95.42	4.58

The effective size, coefficient of uniformity and coefficient of curvature of sand are as listed in table above.

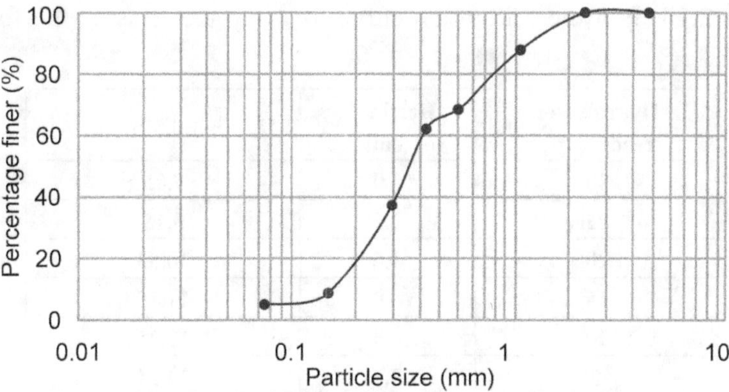

Figure 2.1: Particle size distribution curve for sand.

Question 2.17: A liquid limit test conducted on a soil sample in the cup device gave the following results:

Number of blows	37	28	21	14
Water content (%)	17.27	20.87	25.44	30.69

The determinations of plastic limit gave water content of 16.8%.

Determine liquid limit and plasticity index.

Solution: The observations are plotted as given below:

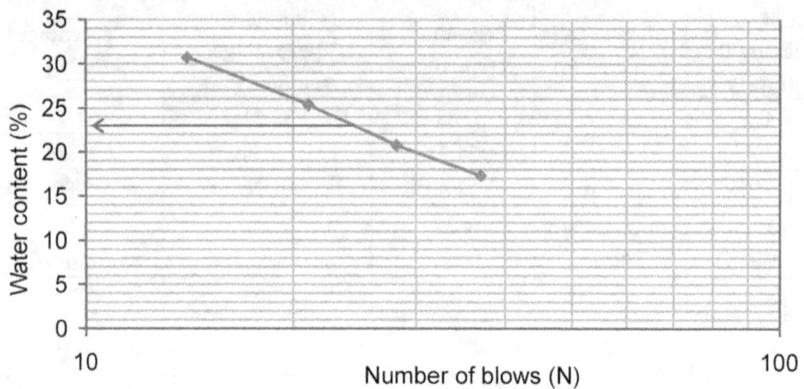

Figure 2.2: Flow curve.

Liquid limit w_L= 23.0 %; Plastic limit w_p = 16.8%;

Plasticity index = $w_L - w_p$ = 23.0 − 16.8 = 6.2%.

Question 2.18: Plot the calibration curve for hydrometer using the following data:

Hydrometer reading, R_h	Height, H (cm)	$H_e = H + 0.5(h − V_h/A)$
30	0.9	7.85
25	2.2	9.15
20	3.6	10.55
15	4.9	11.85
10	6.3	13.25
5	7.6	14.55
0	8.9	15.85
-5	10.3	17.25

Volume of hydrometer, V_h = 100 cm^3.

Height of bulb, h = 16.5 cm.

Cross-sectional area of jar A = 38.46 cm^2.

V_h/A = 2.60 cm, 0.5(h − V_h/A) = 6.95 cm.

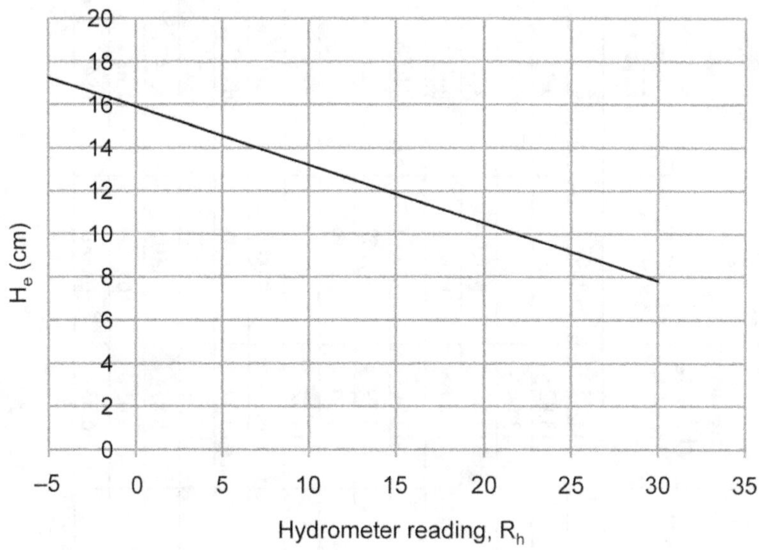

Figure 2.3

Question 2.19: The hydrometer test conducted on a soil sample gave the following results:

Time (minutes)	0	0.375	0.5	1	2	4	8	15	30	60	120	240	1440	2880	4320
Hydrometer reading		42	40.75	39.5	38.25	37	36	34	32	28	23.5	18	10	4.5	2
Effective height (cm)	13.75	13.82	13.89	13.96	14.03	14.08	14.19	14.3	14.52	14.77	15.07	15.51	15.81	15.95	16.03

Composite correction = + 0.5. Plot the grain size distribution curve. $G = 2.54$; $M_d = 70.58$ g.

Solution: The calculations are given in table below:

Using equation, $D = \left\{ \dfrac{3 \times 10^3 (\eta H_e)}{9.81\,(G-1)t} \right\}^{0.5}$; determine particle size D (mm) as given in table.

Use the equation $N' = \dfrac{100\,GR_C}{M_D\,(G-1)}$; for calculating percentage finer as given in table.

The observations are plotted as given below:

Time, t (minutes)	Hydrometer reading, R_h	Temperature	Composite correction, C_m (+)	Corrected hydrometer reading, $R_c = R_h + C_m$	Viscosity, $\eta = 0.00855 \times 10^{-4} \text{kNs/m}^2$	Effective height, H_e (cm)	Particle size, D (mm) $D = \left\{ \dfrac{3 \times 10^3 (\eta H_e)}{9.81 (G-1)t} \right\}^{0.5}$	Percentage finer $N' = \dfrac{100\, GR_C}{M_d (G-1)}$
0						13.75	0.075	100.0
0.375	42.00	27 °C	0.5	42.50		13.82	0.0741	99.31
0.5	40.75		0.5	41.25		13.89	0.0643	96.39
1	39.50		0.5	40.00		13.96	0.0456	93.47
2	38.25		0.5	38.75		14.03	0.0323	90.55
4	37.00		0.5	37.50		14.08	0.0229	87.63
8	36.00		0.5	36.50		14.19	0.0162	85.29
15	34.00		0.5	34.50		14.30	0.0119	80.62
30	32.00		0.5	32.50		14.52	0.0085	75.94
60	28.00		0.5	28.50		14.77	0.0060	66.60
120	23.50		0.5	24.00		15.07	0.0043	56.08
240	18.00		0.5	18.50		15.51	0.0031	43.23
1440	10.00		0.5	10.50		15.81	0.0013	24.54
2880	4.50		0.5	5.00		15.95	0.0009	11.68
4320	2.00		0.5	2.50		16.03	0.0007	5.84

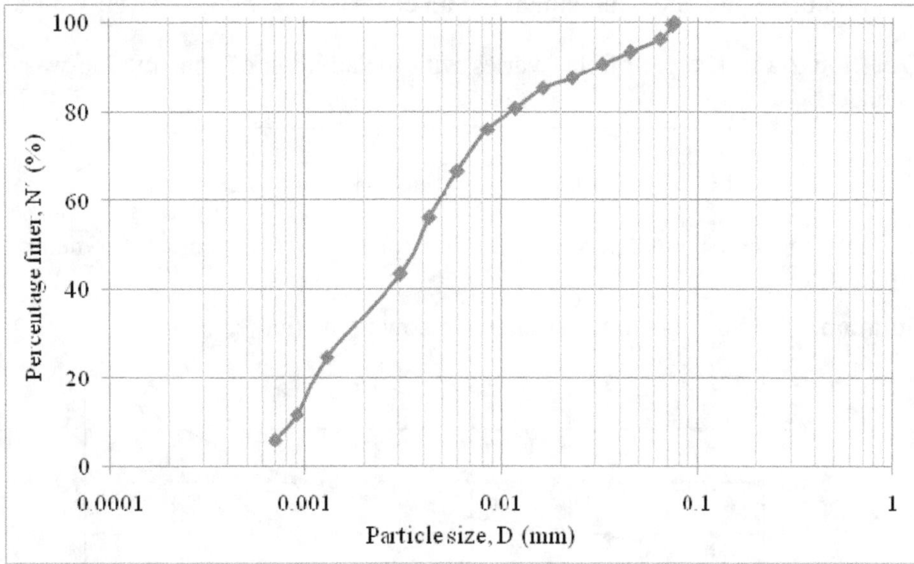

Figure 2.4: Particle size distribution curve using hydrometer analysis.

Question 2.20: The differential free swell of clay varies with the addition of sand as under:

Sand (%)	0	10	20	30	40
Differential free swell (%)	35	16	4	0	0

Plot the curve and determine the minimum optimum content of sand for no swell.

Solution: The curve is plotted as shown below in the figure.

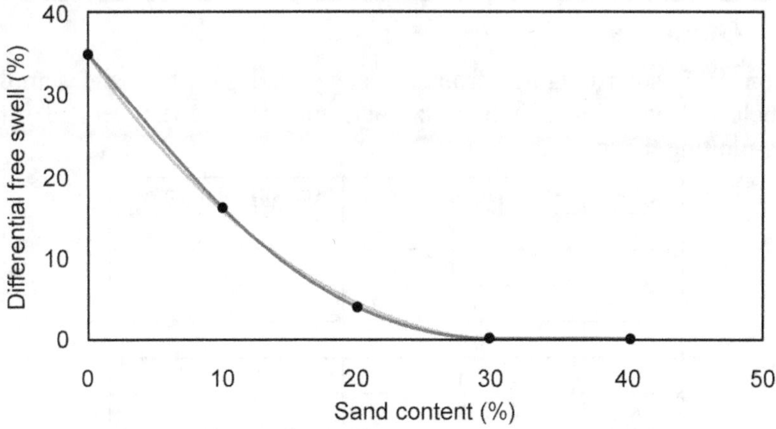

Figure 2.5: Differential free swell versus sand content.

From the figure, the minimum optimum content of sand for no swell = 30%.

Question 2.21: The pH of clay varies with the addition of construction waste as under:

Sand (%)	0	8	16	24	32
pH	6.70	6.79	6.88	7.00	7.28

Plot the curve and determine the construction waste content for neutral mixture.

Solution: The curve is plotted as shown below in the figure.

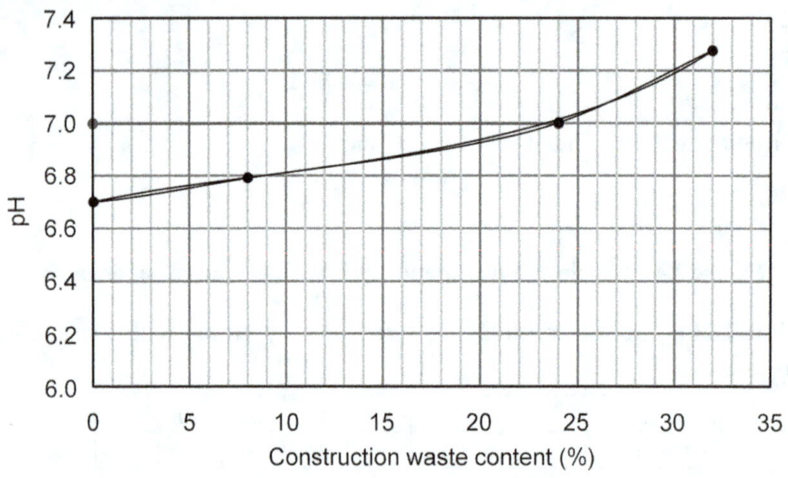

Figure 2.6: pH versus construction waste content.

From the figure, the construction waste content for neutral mixture = 24%.

 **Questions**

Question 1: Results obtained from the sieve analysis of a sand sample are given below. Determine effective size and uniformity coefficient of sand by suitably plotting the results.

S. No.	Mesh opening (mm)	Weight retained (g)
1.	2.40	0.0
2.	1.20	0.0
3.	0.60	30.0
4.	0.30	215.0
5.	0.15	225.0
6.	0.075	25.0
7.	Pan	0.5

(ESE: 1998)

Question 2: Explain and discuss the use of liquidity index, activity number, thixotropy and sensitivity of soil.

(ESE: 1996)

Question 3: What do you understand by index properties of soil? Explain and list the properties under different categories.

(ESE: 2012)

Question 4: A liquid limit test conducted on a soil sample in the cup device gave the following results:

Number of blows	10	19	23	27	40
Water content (%)	60.0	45.2	39.8	36.5	25.2

Two determinations of plastic limit gave water content of 20.3% and 20.8%. Determine:

1. liquid limit and plastic limit
2. plasticity index
3. liquidity index, if the natural water content is 27.4% and
4. void ratio at the liquid limit, if the specific gravity $G_s = 2.7$.

If the soil were to be loaded to failure, would you expect a brittle failure? (ESE: 2013)

Question 5: The plastic limit and liquid limit of soil are 28% and 44% respectively. The percentage volume change from liquid limit to dry state is 40% of the dry volume. Similarly the percentage volume change from plastic limit to dry state is 20% of the dry volume. Determine the shrinkage ratio.

Question 6: A specimen of clayey silt contains 67% silt size particles. Its liquid limit = 35% and plastic limit = 25%. In liquid limit test, at moisture content of 30%, required number of blows was 45. Determine its plasticity index, activity and consistency index.

Question 7: Consider the following properties for clays X and Y:

S. No.	Properties	Clay X (%)	Clay Y (%)
1.	Liquid limit	35	50
2.	Plastic limit	18	32
3.	Natural water content	20	40

Which of the clay, X and Y, experiences large settlement under identical loads; is more plastic and is softer in consistency?

Question 8: A saturated specimen of clay was immersed in mercury and displaced volume was 18.8 cm^3. The weight of sample was 30.2 g. After oven drying for 48 hours, weight reduced to 19.2 g while volume came down to 10.6 cm^3. Determine the shrinkage limit.

Question 9: The liquid limit and plastic limit of sample are 55% and 65% respectively. The percentage of soil fraction with grain size finer than 0.002 mm is 28. Determine the activity ratio of soil sample.

Question 10: A liquid limit test conducted on a soil sample in the cup device gave the following results:

No. of blows	10	19	23	27	40
Water content %	60.00	45.20	39.80	36.50	25.20

Two determinations for the plastic limit gave water content of 20.30% and 20.80%. Determine:

(i)   The liquid limit and plastic limit

(ii) The plasticity index.

(iii) The liquidity index if natural water content is 27.40%

(iv) the void ratio at the liquid limit, if specific gravity = 2.7. If the soil were to be loaded to failure, would you expect a brittle failure?

CHAPTER 3

Soil Classification

Question 3.1: Sieve analysis on a dry soil sample of mass 1000 g showed that 980 g and 270 g of soil pass through 4.75 mm and 0.075 mm sieve, respectively. The liquid limit and plastic limits of the soil fraction passing through 425 m sieves are 40% and 18% respectively. Classify the soil.

Solution: Plasticity index, $PI = w_l - w_p$

$$= 40 - 18 = 22\%$$

Since PI > 17, soil is highly plastic i.e., clayey soil.

Since, 98% of soil passes through 4.75 mm sieve and 27% through 0.075 mm sieve, majority (more than 50%) of soil lies between 0.075 and 4.75 mm i.e., sand.

∴ Soil is clayey sand (SC)

Question 3.2: A soil has a liquid limit of 45% and lies above A-line when plotted on a plasticity chart. Determine the group symbol of the soil as per IS soil classification.

Solution: Liquid limit of soil is 45% which lies between 35% - 50%; hence the soil is of intermediate plasticity. Since the soil lies above A-line, the symbol should be CL.

Question 3.3: If the proportion of soil passing 75 micron sieve is 50% and liquid limit and plastic limit of soil are 40% and 20% respectively, determine the group index of the soil.

Solution: Group index, $GI = 0.2a + 0.005ac + 0.01bd$

a = percentage passing 75 mm sieve greater than 35% but not exceeding 75% (between 0 to 40)

b = percentage passing 75 mm sieve greater than 15% but not exceeding 55% (between 0 to 40)

c = liquid limit greater than 40% but not exceeding 60% (between 0 to 20)

d = plasticity index greater than 10 and not exceeding 30 (between 0 to 20)

$$a = 50 - 35 = 15 < 40$$

$$b = 50 - 15 = 35 < 40$$

$$c = 40 - 40 = 0$$

$$d = 20 - 10 = 10 < 20$$

$$\therefore \; GI = 0.2 \times 15 + 0.005 \times 15 \times 0 + 0.01 \times 35 \times 10$$

$$= 3 + 0 + 3.5$$

$$= 6.5$$

Question 3.4: Based on grain distribution analysis, the D_{10}, D_{30} and D_{60} values of a given soil are 0.23 mm, 0.30 mm and 0.41 mm respectively. Classify the soil classification as per Indian standard.

Solution: Coefficient of uniformity,

$$C_u = D_{60}/D_{10}$$

$$= 0.41/0.23 = 1.78$$

Coefficient of curvature,

$$C_c = (D30)^2/(D_{60} \times D_{10})$$

$$= 0.30^2/ (0.41 \times 0.23) = 0.95$$

For well graded sand $1 < C_c < 3$ and $C_u > 6$ as per IS classification. So the soil is poorly graded sand.

Question 3.5: Given that the coefficient of curvature = 1.4; $D_{30} = 3$ mm, $D_{10} = 0.6$ mm. Based on this information of particle size distribution for use as subgrade, classify the soil.

Solution: Coefficient of curvature,

$$C_c = D_{30}^2/(D_{60} \times D_{10})$$

$$D_{60} = D_{30}^2/(C_c \times D_{10})$$

Coefficient of uniformity,

$$C_u = D_{60}/D_{10}$$

$$= D_{30}^2/(C_c \times D_{10}^2)$$

$$= [3/0.6]^2 \times 1/1.4 = 17.9$$

Since $1 < C_c < 3$ and $C_u > 6$, hence the soil is well graded sand.

Question 3.6: On the analysis of particle size distribution of soil, it is found that $D_{10} = 0.1$ mm, $D_{30} = 0.3$ mm and $D_{60} = 0.8$ mm. Determine the coefficient of uniformity and coefficient of curvature of the soil.

Solution: Coefficient of uniformity,

$$C_u = D_{60}/D_{10}$$

$$= 0.8/0.1 = 8$$

Coefficient of curvature,

$$C_c = D_{30}{}^2 / (D_{60} \times D_{10})$$
$$= 0.3^2 / (0.8 \times 0.1) = 1.125.$$

Question 3.7: The liquid limit and plastic limit values of different soil samples are given below:

Liquid limit (%)	60	45	44	22	56	33	31	25
Plastic limit (%)	25	33	21	19	35	26	19	19

Plot the values on A - line and classify the soil samples.

Solution: The calculations are given in table below:

Liquid limit (%)	60	45	44	22	56	33	31	25
Plastic limit (%)	25	33	21	19	35	26	19	19
Plasticity index (%)	35	12	23	3	21	7	12	6
Classification	CH	MI or OI	CI	ML	MH or OH	ML or OL	CL	CL or ML

The values are plotted in the A- line chart and the soil samples are classified as shown below:

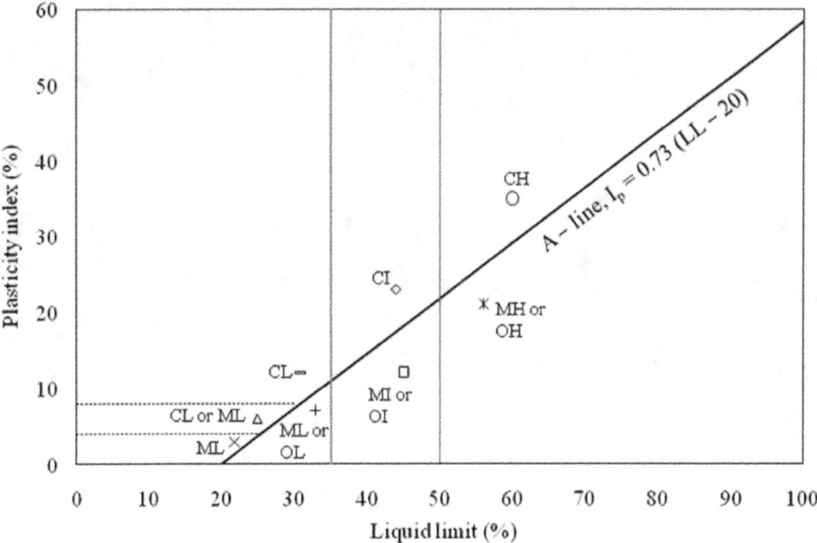

Figure 3.1: A – line showing the soil classification.

 ## Questions

Question 1: If the proportion of soil passing 75micron sieve is 55% and liquid limit and plastic limit are 35% and 20% respectively. Determine the group index of the soil.

Question 2: On the analysis of particle size distribution of soil, it is found that $D_{10} = 0.6$ mm, $D_{30} = 0.9$ mm and $D_{60} = 1.2$ mm. Determine the coefficient of uniformity and coefficient of curvature as given by the particle size distribution curve.

Question 3: Given that the coefficient of curvature = 2, $D_{30} = 6$ mm and $D_{10} = 2$mm.Based on this information of particle size distribution for use as subgrade, classify the soil.

Question 4: Based on grain distribution analysis, the D_{10}, D_{30} and D_{60} values of a given soil are 0.63 mm, 0.84 mm and 1.30 mm respectively. Classify the soil as per Indian standard.

Question 5: A soil has a liquid limit of 55% and lies above A-line when plotted on a plasticity chart. Determine the group symbol of the soil as per IS soil classification.

Question 6: Sieve analysis on a dry soil sample of mass 1000g showed that 980g and 270g of soil passthrough 4.75mm and 0.075mm sieve, respectively. The liquid limit and plastic limits of the soil fraction passing through 425μ sieves are 40% and 18%respectively. Classify the soil.

CHAPTER 4

Permeability of Soil

Question 4.1: A capillary permeability test was conducted in two-stage under a head of 60 cm and 180 cm respectively at the entry end. In the first stage, the wetted surfaced moved from 1.5 cm to 7 cm in 7 minutes. In the second stage, it advanced from 7 cm to 18.5 cm in 24 minutes. The degree of saturation at the end of the test was 85% and the porosity was 35%. Determine the capillary head and the co-efficient of permeability.

Solution: As we know that, the capillarity-permeability test,

The expression $\dfrac{x_2^2 - x_1^2}{t_2 - t_1} = \dfrac{2k_u}{S\,n}(h_0 + h_c)$

First stage, $\qquad \dfrac{7^2 - 1.5^2}{7} = \dfrac{2k_u}{0.85 \times 0.35}(60 + h_c)$

or $\qquad k_u(60 + h_c) = 0.9934$

Second stage, $\qquad \dfrac{18.5^2 - T^2}{18.5} = \dfrac{2k_u}{0.85 \times 0.35}(180 + h_c)$

or $\qquad k_u(180 + h_c) = 1.8175$

From equations (i) and (ii), we get $\dfrac{180 + h_c}{60 + h_c} = \dfrac{1.8175}{0.9934} = 1.8296$,

From which, $h_c = 84.65$ cm

Hence from equation (i)

$$k_u = 6.87 \times 10^{-3} \text{ cm/min.} = 1.145 \times 10^{-4} \text{ cm/sec.}$$

So, the capillarity head is 84.65 cm

And the co-efficient of permeability is 1.145×10^{-4} cm/sec.

Question 4.2: What will be the ratio of average permeability in horizontal direction to that in the vertical direction for a soil deposit consisting of three horizontal layers, if the thickness and permeability of the second layer are twice of those of the first and those of the third layer twice those of second?
Solution: Let k_1 and z_1 stands for first layer.

For second layer, $\qquad k_2 = 2k_1$ and $z_2 = 2z_1$

For third layer, $\qquad k_3 = 2k_2 = 4k_1$ and $z_2 = 2z_2 = 4z_1$

Total

$$z = z_1 + z_2 + z_3$$
$$= z_1 + 2z_1 + 4z_1 = 7z_1$$

Therefore,

$$k_H = (k_1z_1 + k_2z_2 + k_3z_3)/z$$
$$= (k_1z_1 + (2k_1)(2z_1) + (4k_1)(4z_1))/7z_1$$
$$= (21/7)\, k_1$$

$$k_V = z/((z_1/k_1) + (z_2/k_2) + (z_3/k_3))$$
$$= 7z_1/((z_1/k_1) + (2z_1/2k_1) + (4z_1/4k_1))$$
$$= (7/3)\, k_1$$

So, the average permeability in horizontal direction to that in the vertical direction for a soil deposit consisting of three horizontal layers is

$$k_H/k_V = (21 \times 3)/(7 \times 7) = 9/7$$

Question 4.3: A non-homogeneous soil deposit consists of a silt layer sandwiched between a fine sand layer at top and a clay layer below. Permeability of the silt layer is 10 times the permeability of the clay layer and one-tenth of sand layer. Thickness of silt layer is 2 times the thickness of sand layer and two-third of the thickness of clay layer. Determine the ratio of equivalent horizontal and equivalent vertical permeability of deposit.

Solution:

$$k_1 = 10k_2 = 100\, k_3$$
$$k_2 = k_1/10$$
$$k_H = (k_1H_1 + k_2H_2 + k_3H_3)/(H_1 + H_2 + H_3)$$
$$= 100k_3H_1 + 10k_3 \times 2H_1 + k_3 \times 3H_1)/(H_1 + 2H_1 + 3H_1)$$
$$= 20.5\, k_3$$
$$k_V = (H_1 + H_2 + H_3)/(H_1/k_1 + H_2/k_2 + H_3/k_3)$$
$$= (H_1 + 2H_1 + 3H_1)/(H_1/100k_3 + 2H1/10k_3 + 3H_1/k_3)$$
$$= 1.869\, k_3$$
$$= k_H/k_V = 20.5\, k_3/1.869\, k_3 = 10.9684$$

Question 4.4: A 30 cm well completely penetrates an unconfined aquifer of depth 40m. After a long period of pumping at a steady rate of 1500 litres/minute, the drawdown in two observations wells 25 m and 75 m away from the pumping well were found to be 3.5 m and 2.0 m respectively. Determine the transmissibility of the aquifer. What is the drawdown at the pumping well?

Solution: $q = 1500$ litres/min $= 1.5$ m^3/min $= 0.25$ m^3/sec

$$h_2 = H - s_2 = 40 - 2 = 38 \text{ m}; \quad h_1 = H - s_1 = 40 - 3.5 = 36.5 \text{ m}$$

$$q = \frac{1.36\,k\,(h_2^2 - h_1^2)}{\log_{10}(r_2/r_1)}$$

or $\qquad 0.025 = \dfrac{1.36\,k\,(38^2 - 36.5^2)}{\log_{10}(75/25)}$

From which

$$k = 7.848 \times 10^{-5}\,\text{m/sec}$$

$$T = H\,k = 40 \times 7.848 \times 10^{-5}$$

$$= 3.14 \times 10^{-3}\,\text{m}^2/\text{sec}$$

Also, at the well face,

$$q = \frac{1.36\,k\,(h_1^2 - h_w^2)}{\log_{10}(r_1/r_w)}$$

$$h_w^2 = 36.5^2 - 520.42 = 811.83$$

From which $h_w = 28.49$ m

Draw down at well face, $s_w = H - h_w = 40 - 28.49$

$$= 11.51\,\text{m}$$

Question 4.5: In a falling head permeameter test on a silty clay sample, the following results were obtained: sample length 120 mm; sample diameter 80 mm; initial head 1200mm; final head 400 mm; time for fall in head 6 minutes and stand pipe diameter 4 mm. Find the coefficient of permeability of soil in mm/sec.

Solution: Given,

Sample length, L = 120 mm;

Time for fall in head, t = 6 min = 360 sec;

Initial head, h_1 = 1200 mm;

Final head, h_2 = 400 mm;

Sample diameter = 80 mm;

Sample area, $A = \dfrac{\pi}{4}(80)^2 = 5026.5$ mm^2;

Stand pipe diameter = 4 mm;

Stand pipe area, $a = \dfrac{\pi}{4}(4)^2 = 12.57$ mm^2

In falling head permeameter test, coefficient of permeability of soil is given by

$$k = 2.303\,\frac{aL}{At}\,log_{10}\frac{h_1}{h_2}$$

$$k = 2.303\,\frac{12.57 \times 120}{5026.5 \times 360}\,log_{10}\frac{1200}{400}$$

$$k = 9.159 \times 10^{-4}\,\text{mm/sec}$$

Question 4.6: In a consolidation test, the void ratio of the specimen which was 1.068 under the effective pressure of 214 kN/m^2, changed to 0.994 when the pressure was increased to 429 kN/m^2. Calculate the coefficient of compressibility, compression index and the coefficient of volume compressibility. Find the settlement of foundation resting on above type of clay, if thickness of layer is 8 m and the increase in pressure is 10 kN/m^2.

Solution: Given,

$$e_0 = 1.068; \; e = 0.994; \; \sigma_0' = 214 \text{ kN/m}^2; \; \sigma' = 429 \text{ kN/m}^2$$

Now, coefficient of compressibility, $a_v = \dfrac{\Delta e}{\Delta \sigma'} = \dfrac{e_0 - e}{\sigma' - \sigma_0'}$

$$a_v = \frac{1.068 - 0.994}{429 - 214} = 3.442 \times 10^{-4} \text{ m}^2/\text{kN}$$

Compression Index, $C_c = \dfrac{\Delta e}{\log_{10} \dfrac{\sigma'}{\sigma_0'}}$

$$C_c = \frac{1.068 - 0.994}{\log_{10} \dfrac{429}{214}} = 0.245$$

Coefficient of volume compressibility, $m_v = \dfrac{a_v}{1 + e_0}$

$$m_v = \frac{3.442 \times 10^{-4}}{1 + 1.068} = 1.664 \times 10^{-4} \text{ m}^2/\text{kN} \text{ and}$$

Settlement of foundation, $\Delta H = \dfrac{C_c H_0}{1 + e_0} \log_{10} \dfrac{\sigma_0' + \Delta \sigma}{\sigma_0'}$

$$\Delta H = \frac{0.245 \times 8}{1 + 1.068} \log_{10} \frac{214 + 10}{214} = 0.0188 \text{ m} = 18.8 \text{ mm.}$$

Question 4.7: The compression curve (void ratio, e vs. effective stress, σ'_v) for a certain clayey soil is a straight line in a semi-logarithmic plot and it passes through the points (e =1.2; σ'_v =50 kPa) and (e =0.6: σ'_v = 800 kPa). Determine the compression index of the soil.

Solution: The compression curve for a clayey soil is shown in figure (Paint)

Compression index

$$c_c = \frac{\Delta e}{\log_0 \left[\frac{\sigma_2}{\sigma_1}\right]} = \frac{1.2 - 0.6}{\log 10 \left[\frac{800}{50}\right]} = \frac{0.6}{\log_{10} 16} = 0.498$$

Question 4.8: If during a permeability test on a soil sample with a falling head permeameter, equal time intervals are noted for drop of head from 'h_1' to 'h_2' and again from 'h_2' to 'h_3', then which one of the following relations would hold good.

Solution: In falling head permeability test,

$$\text{Permeability,} \quad k = 2.3 \, (aL/At) \, \log(h_1/h_2)$$

$$= 2.3 \, (aL/At) \, \log(h_2/h_3)$$

$h_1/h_2 = h_2/h_3$

$\Rightarrow \quad h_2^2 = h_1/h_3$

Question 4.9: A bed of sand consists of three horizontal layers of equal thickness. The value of Darcy's 'k' for the upper and lower layers is 1×10^{-2} cm/sec and that for the middle layer is 1×10^{-1} cm/sec. Determine the ratio of the permeability of the bed in the horizontal direction to that in the vertical direction.

Solution: $k_{ex}/k_{ez} = [(k_1 h_1 + k_2 h_2 + k_3 h_3)/(h_1 + h_2 + h_3)^2] \times (h_1/k_1 + h_2/k_2 + h_3/k_3)$

For $h_1 = h_2 = h_3 = 1$ (say)

$k_{ex}/k_{ez} = (k_1 + k_2 + k_3)/9(1/k_1 + 1/k_2 + 1/k_3)$

$k_1 = k_3 = 1 \times 10^{-2}$ cm/s

$k_2 = 10 \times 10^{-2}$ cm/s

$k_{ex}/k_{ez} = [(1 + 10 + 1)/9] \times (1/1 + 1/10 + 1/1)$

$= 12 \times 2.1/9 = 2.8$

Question 4.10: Due to rise in temperature, the viscosity and unit weight of percolating fluid are reduced to 70% and 90% respectively. Other things being constant, the change in coefficient of permeability will be?

Solution: Taylor's equation based on Poiseuille's law for laminar flow through circular tube is given by

$k = [\gamma e^3/ \mu (1+e)] \times CD_s^2$

Change in coefficient of permeability,

$k' = [0.9\gamma e^3/ 0.7\mu (1+e)] \times CD_s^2$

$k'/k = (0.9/0.7) \times 100 = 128.57$

Change $= 28.57\%$

Question 4.11: A soil has discharge velocity of 5×10^{-7} m/s and a void ratio of 0.50. Determine the seepage velocity.

Solution: Porosity is expressed as

$n = e/1+e = 0.5/(1+0.5) = 1/3$

Seepage velocity,

$v_s = v/n = 5 \times 10^{-7}/(1/3)$

$= 15 \times 10^{-7}$ m/s

Question 4.12: Two observations wells penetrating into a confined aquifer are located 1500 m apart in the direction of flow. Heads of 50 m and 25 m are indicated at these two observation wells. If the coefficient of permeability for the aquifer is 30 m/day and its porosity is 0.25, determine the time of travel of an inert tracer from one well to another.

Solution: Let the inert tracer travel by a distance 'dx' in time 'dt'

v_s = dx/dt = Actual velocity of flow

v_s = v/n = dx/dt

Now,   v = ki

$\Rightarrow$ v_sn = ki

$\Rightarrow$ n dx/ dt = ki

$\Rightarrow$ (1500/t) 0.25 = 30 × (50 – 25)/1500

$\Rightarrow$ t = 750 days

Question 4.13: The void ratio of a given soil A is twice that of another soil B, while the effective size of particles of soil A is 1/3 of that of soil B. Determine the ratio of height of capillary rise of water in soil A to that in soil B.

Solution: The empirical formula for capillary height is

$h = C/eD_{10}$

$\Rightarrow$ $h_A/h_B = e_B(D_{10})_B/ e_A(D_{10})_A = (e_B/e_A) × (D_{10B}/D_{10A})$

$= \frac{1}{2} ×3 = 1.5$

Question 4.14: A stratum of soil consists of three layers of equal thickness. The permeability of both the top and bottom layers is 10^{-4} cm/sec; and that of the middle layer is 10^{-3} cm/sec. Determine the value of horizontal coefficient of permeability for the entire composite of the soil layer.

Solution: $k_H = [k_1H + k_2H + k_3H]/ 3H = [10^{-4} + 10^{-3} + 10^{-4}]/3$

$= 12 × 10^{-4}/3 = 4 × 10^{-4}$ cm/sec

Question 4.15: An extended layer of soil with homogeneous rounded grains has 10% of the material finer than 0.07 mm. The constant to be adopted to determine its permeability has been recommended as 750. What is the permeability?

Solution: $k = CD_{10}^2$

$k = 750 × (0.07)^2 = 3.675$ m/day

As per Allen Hazen

$k = CD_{10}^2$

Where   c = 100

D_{10} is in cm

And k is in cm/sec

Similarly, if we solve the equation

Then $k = 750 \times (0.007 \text{cm})^2$

$\quad = 0.03675 \text{ cm/sec}$

$\quad = 0.03675 \times 10^{-2} \text{ m/day} \times 24 \times 3600$

$\quad = 31.752 \text{ m/day}$

Question 4.16: Determine the permeability of soil using constant head permeability test from the following data:

Diameter of pipe d = 5 cm

Diameter of cylindrical soil sample D = 8 cm

Height of soil specimen L = 6 cm

Discharge Q = 560 cm^3

Hydraulic head h =155 cm

Time interval t = 30 seconds

Solution:

Cross sectional area of stand pipe a = π = $\pi \times (5^2/4) = 19.62 \text{ cm}^2$

Cross sectional area of soil specimen A = $\pi \times (D^2/4) = \pi \times (8^2/4) = 50.28 \text{ cm}^2$

Coefficient of permeability $k = \dfrac{QL}{hAT} = \dfrac{560 \times 6}{50.28 \times 155 \times 30} = 0.01437 \text{cm/sec}$

$\quad = 1.437 \times 10^{-2} \text{cm/sec.}$

Questions

Question 1: A laboratory constant head permeability test was conducted on a silty sand specimen of void ratio 0.45. The cylindrical specimen had a diameter of 7.3 cm and a height of 16.8 cm. The head during the test was 75 cm. After 1 minute of testing at room temperature of $20\degree$C, a total 775.6 gm of water was collected. Compute the coefficient of permeability in metres per sec. If the void ratio changes to 0.38, what would be the change in permeability?

(ESE: 2003)

Question 2: Estimate the value of coefficient of permeability for a uniform graded sand of size $D_{10} = 0.15$ mm obtained from sieve analysis. G = 2.67.

(ESE: 2010)

Question 3: What is the intrinsic permeability of a water saturated medium that has hydraulic conductivity of 15.24 m/day? Assume the ground water is at atmospheric pressure at 20°C and has a density of 998.2 kg/m^3 and viscosity of 1.002×10^{-3} kg/m s.

(ESE: 2011)

Question 4: A horizontal stratified deposits consists of four layers each uniform in itself. The permeabilities of the layers are 7.5×10^{-4} cm/sec, 49×10^{-4} cm/sec, 13×10^{-4} cm/sec and 17×10^{-4} cm/sec and their thickness are 5 m, 4 m, 17 m, and 6 m respectively. Find the effective average permeabilities of the deposit in horizontal and vertical directions.

(ESE: 2012)

Question 5: The compression curve (void ratio, e vs. effective stress, σ'_v) for a certain clayey soil is a straight line in a semi-logarithmic plot and it passes through the points (e = 1.4; σ'_v = 40 kPa) and (e = 0.8: σ'_v = 600 kPa). Determine the compression index of the soil.

Question 6: A capillary permeability test was conducted in two stages under a head of 70 cm and 190 cm respectively at the entry end. In the first stage, the wetted surfaced moved from 1.8 cm to 8 cm in 10 minutes. In the second stage, it advanced from 9 cm to 18.5 cm in 30 minutes. The degree of saturation at the end of the test was 90% and the porosity was 45%. Determine the capillary head and the co-efficient of permeability.

Question 7: An extended layer of soil with homogeneous rounded grains has 15% of the material finer than 0.09 mm. The constant to be adopted to determine its permeability has been recommended as 850. What is the co-efficient of permeability?

Question 8: Two observations wells penetrating into a confined aquifer are located 1600 m apart in the direction of flow. Heads of 60 m and 30 m are indicated at these two observation wells. If the coefficient of permeability for the aquifer is 40 m/day and its porosity is 0.35, the time of travel of an inert tracer from one well to another is?

Question 9: A soil has discharge velocity of 6×10^{-7} m/s and a void ratio of 0.80. Determine it seepage velocity.

Question 10: A non-homogenous soil deposit consists of a silt layer sandwiched between a fine-sand layer at top and a clay layer below. Permeability of the silt layer is 10 times the permeability of the clay layer and one-tenth of the permeability of the sand layer. Thickness of the silt layer is twice the thickness of the sand layer and two-third of the thickness of the clay layer. Determine the ratio of equivalent horizontal and equivalent vertical permeability of the deposit.

Question 11: The coefficients of permeability of soil in horizontal and vertical directions are 3.46 and 1.5 m/day respectively. The base length of a concrete dam resting in this soil is 100 m. When the flow net is developed for this soil with 1:25 scale factor in the vertical direction? Determine the reduced base length of the dam.

Question 12: In a falling head permeability test of the initial head of 1.0 m dropped to 0.35 m in 3 hours, the soil specimen was 200 mm long and of 100 mm diameter. Determine the coefficient of permeability of the soil.

Question 13: Two soil specimens with identical geometric dimensions were subjected to falling head permeability tests in the laboratory under identical conditions. The fall of water head was measured after an identical time interval. The ratio of initial to final water heads for the test involving the first specimen was 1.25. If the coefficient of permeability of the second specimen is 5 times that of the first, determine the ratio of initial to final heads in the test involving the second specimen.

Question 14: In a falling head permeability test on a 120 mm high and 100 mm diameter cylindrical sample, the water level in the standpipe dropped from a height of 750 mm to 250 mm in one hour. Determine the coefficient of permeability. Take the internal diameter of the standpipe equal to 6 mm.

Question 15: How the average permeability of a soil deposit consisting of a number of layers is determined?

Question 16: The soil at a site consists of two layers of thickness H each. The coefficient of permeability of the soil of 1st layer is k_1 in both horizontal and vertical directions, whereas for the 2nd layer, it is $k_1/2$. What will be the equivalent permeability of the two-layered soil in horizontal and vertical directions?

Question 17: 0.5 m diameter well fully penetrates an unconfined aquifer whose bottom is 150m below the undisturbed groundwater table. The water is pumped out from the well at a constant rate of $0·1$ m^3/s. After equilibrium is reached, the drawdown observed in two observation wells at radial distances of 10 m and 50 m are respectively 10 m and 5 m. Determine (i) the coefficient of permeability (in m/s) (ii) the drawdown in the well (in m)

CHAPTER 5

Seepage Analysis

Question 5.1: A uniform collapsible sand stratum, 2.5 m thick, has specific gravity of sand as 2.65, with a natural void ratio of 0.65. Determine the hydraulic head required to cause quick collapsible sand condition.

Solution: $i_{cr} = (G - 1)/ (1+e) = 1.65/1.65 = 1$.

$i_{cr} = h/H$

$h = i_{cr} \times H$

$\quad = 1 \times 2.5$

$\quad = 2.5m$

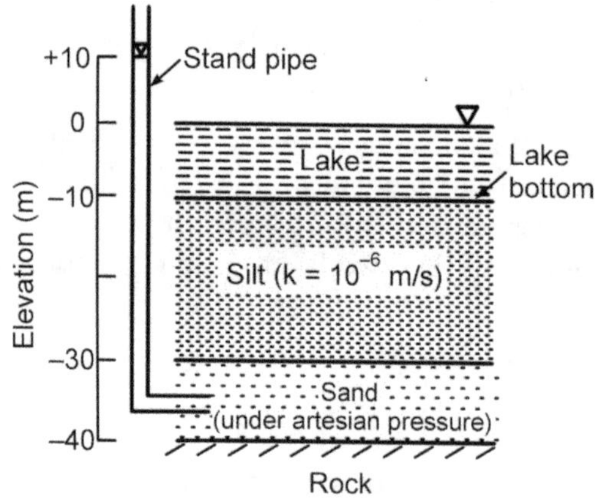

Figure 5.1: Soil profile below a lake.

Question 5.2: The soil profile below a lake with water level at elevation $= 0m$ and lake bottom at elevation $= -10m$ is shown in the figure 5.1, where k is the permeability coefficient. A piezometer (stand pipe) installed in the sand layer shows a reading of +10 m elevation. Assume that the piezometric head is uniform in the sand layer. Determine the quantity of water (in m^3/s) flowing into the lake from the sand layer through the silt layer per unit area of the lake bed.

Solution: From the diagram,10 m of head is lost in 20 m thickness of silt layer before reaching the lake

$$i = h/z = 10/20 = 0.5$$

$$Q = kiA = 10^{-6} \times 0.5 \times 1$$

$$= 0.5 \times 10^{-6} \text{ m/s}$$

Question 5.3: A 0.80 m deep bed of sand filter (length 4m and width 3m) is made of uniform particles (diameter = 0.40 mm, specific gravity = 2.65, shape factor = 0.85) with bed porosity of 0.4. The bed has to be backwashed at a flow rate of 3.60 m^3/min. During backwashing, if the terminal settling velocity of sand particles is 0.05 m/s, determine the expanded bed depth.

Solution: Expanded bed depth $= D_e = D \left[\dfrac{1-n}{1-n_e} \right]$

$D =$ depth of bed $= 0.8$m; n= bed porosity $= 0.4$

$n =$ porosity of expanded bed $= \left(\dfrac{v_b}{v_s} \right)^{0.22}$

$v_s =$ terminal velocity of settling particles during back wasting $= 0.05$ m/s

$v_b =$ backwash velocity

$\quad = \dfrac{Q}{BL}$ (Q = backwash flow rate)

$Q = 3.6$ m^3/min, B $= 3$m and L$=4$ m

$\Rightarrow v_b = \dfrac{3.6}{60 \times 3 \times 4} = 0.005$ m/s

$\Rightarrow n_e = \left(\dfrac{0.005}{0.05} \right)^{0.22} = 0.60255$

$\therefore D_e = 0.8 \left[\dfrac{1 - 0.4}{1 - 0.60255} \right]$

$\quad = 1.21$ m/s

Question 5.4: A uniform sand stratum 2.5 m thick has a specific gravity of 2.62 and a natural void ratio of 0.62. Determine the hydraulic head required to cause quick sand condition in the sand stratum.

Solution: Critical hydraulic gradient

$$i_c = (G - 1) / (1 + e)$$

$$= (2.62 - 1) / (1 + 0.62)$$

$$= 1.0$$

Head required $= i_c \times$ thickness of sand stratum

$$= 1 \times 2.5$$

$$= 2.5 \text{ m}$$

Question 5.5: A soil has a discharge velocity of 6×10^{-7} m/s and a void ratio of 0.5. Determine its seepage velocity.

Solution: Porosity n = e / (1 + e) = 0.5/1.5 = 1/3

Seepage velocity = discharge velocity/ porosity

$$= v/n$$

$$= 6 \times 10^{-7} / (1/3)$$

$$= 18 \times 10^{-7} \text{ m/s}$$

Question 5.6: An upward hydraulic gradient 'i' of a certain magnitude will initiate the phenomenon of boiling in granular soils. Determine the magnitude of this gradient.

Solution: Critical hydraulic gradient,

$$i_{cr} = (G - 1)/ (1 + e)$$

$$G = 2.65, e = 0.7$$

On solving,

$$i_{cr} = 1$$

Question 5.7: Which one of the following equations correctly gives the relationship between the specific gravity of soil grains 'G' and the hydraulic gradient 'i' to initiate 'quick' condition in the sand having a void ratio of 0.5?

Solution: Critical hydraulic gradient,

$$i_c = (G - 1)/ (1 + e)$$

$$= (G - 1)/ (1 + 0.5)$$

$$G = 1.5 \, i_c + 1$$

Question 5.8: The sand deposit has a porosity of 1/3 and its specific gravity is 2.5. Determine the critical hydraulic gradient to cause sand boiling in the stratum.

Solution: $i_c = (G - 1) (1 - n)$

$$= (2.5 - 1) \times (1 - (1/3))$$

$$= 1.0$$

Question 5.9: A flownet, constructed to determine the seepage through an earth dam which is homogeneous but anisotropic, gave four channels and sixteen equipotential drops. The coefficients of permeability in the horizontal and the vertical directions are 4.0×10^{-7} m/s and 1×10^{-7} m/s respectively. If the storage head was 20 m, then determine the seepage per unit length of dam (in m^3/s).

Solution: Seepage is given by

$$q = (k_x k_z)^{1/2} \times (N_f / N_d) \times H$$

Substituting the values

$$= (4 \times 10^{-7} \times 1 \times 10^{-7})^{1/2} \times (4/16) \times 20$$

$$= 10 \times 10^{-7} \ m^3/s/ \ m$$

Question 5.10: A sand deposit has a porosity 0.375 and specific gravity 2.6. Determine the critical hydraulic gradient for sand deposit.

Solution: Critical hydraulic gradient,

$$i_c = (G - 1) / (1 + e)$$

$$= (G - 1) (1 - n)$$

$$= (2.6 - 1) (1 - 0.375)$$

$$= 1.0$$

Question 5.11: A flownet for an earth dam on impervious foundation consists of 4 flow channels and 15 equipotential drops. Full reservoir level is 15 m above the downstream horizontal filter. Given that horizontal permeability is 9 $\times$ 10^{-6} m/s and the vertical permeability is 1 $\times$ 10^{-6} m/s, then determine the quantity of seepage through the dam.

Solution: Equivalent permeability,

$$k_e = (k_x k_z)^{1/2}$$

$$= (9 \times 10^{-6} \times 1 \times 10^{-6})^{1/2}$$

$$= 3 \times 10^{-6}$$

Seepage, $q = k_e \times (N_f / N_d) \times H$

$$= 3 \times 10^{-6} \times (4/ 15) \times 15$$

$$= 12 \times 10^{-6} \ m^3/s/m$$

Question 5.12: A strata of 3.5 m thick fine sand has a void ratio 0.7 and specific gravity 2.7. For quick sand condition to develop in these strata, determine the head required for flow of water in upward direction.

Solution: The critical hydraulic gradient,

$$i_c = (G - 1)/ (1 + e)$$

$$= (2.7 - 1)/ (1 + 0.7)$$

$$= 1.7/1.7 = 1.0$$

For quick sand condition,

$$i = i_c = h/L$$

$$h = 1.0 \times 3.5 = 3.5 \ m$$

Question 5.13: A flownet of a coffer dam foundation has 6 flow channels and 18 equipotential drops. The head of water lost during seepage is 6 m. If the coefficient of permeability of foundation is 4×10^{-5} m/min, then determine the seepage loss per meter length of dam.

Solution: Seepage loss per meter length,

$$q = k \times (N_f/N_d) \times H$$

Given, $k = 4 \times 10^{-5}$ m/ min

$$= 4 \times 10^{-5} \times 24 \times 60 \text{ m/day}$$

$$= 0.0576 \text{ m/day}$$

N_f is the number of flow channels $= 6.0$

N_d is number of equipotential drops $= 18$

$$q = 0.0576 \times (6/18) \times 6$$

$$= 0.1152 \text{ m}^3/\text{day}$$

$$= 11.52 \times 10^{-2} \text{ m}^3/\text{day}$$

Question 5.14: A sample has a bulk unit weight of 20 kN/m^3 and a degree of saturation of 70%. If the specific gravity of soil grains is 2.65, then determine the value of critical hydraulic gradient of soil.

Solution: Hydraulic gradient, $i_c = (G - 1)/ (1 + e)$

Bulk unit weight $\gamma = (G + Se) \, \gamma_w / (1 + e)$

Taking unit weight of water $\gamma_w = 10$ kN/m^3

$$20 = (2.65 + 0.7e) \times 10/ (1 + e)$$

$$20 + 20e = 26.5 + 7e$$

$$6.5 = 13e$$

$$e = 0.5$$

$$i_c = (2.65 - 1) / (1 + 0.5) = 1.65/1.5$$

$$i_c = 1.1$$

Questions

Question 1: A uniform collapsible sand stratum, 2.5 m thick, has specific gravity of sand as 2.65, with a natural void ratio of 0.65. Determine the hydraulic head required to cause quick collapsible condition.

(ESE: 2016)

Question 2: A soil sample has an average grain diameter 0.03 mm. The size of interstices is one-eighth of the mean grain diameter. Considering σ of water as 0.075 g/cm, determine the height of rise of water in the clay.

(ESE: 2018)

Question 3: Masonry is founded on pervious sand. A factor of safety of 4 is required against boiling. For the sand, n = 45% and G_s = 2.65. Determine the maximum permissible upward hydraulic gradient.

(ESE: 2019)

Question 4: A flownet of a coffer dam foundation has 6 flow channels and 18 equipotential drops. The head of water lost during seepage is 8 m. If the coefficient of permeability of foundation soil is 9×10^{-5} m/min, then determine the seepage loss per meter length of dam.

Question 5: A uniform collapsible sand stratum, 5 m thick, has specific gravity of sand as 2.65, with a natural void ratio of 1.65. Determine the hydraulic head required to cause quick collapsible sand condition.

Question 6: What do you understand by seepage pressure and quick sand condition?

(ESE: 2008)

Question 7: A deposit of sand has a porosity of 45%. Estimate the critical hydraulic gradient to develop quick sand condition if the specific gravity of grain is 2.7.

(ESE: 2009)

Question 8: A flownet is plotted for a homogeneous earthen dam of 30 m height with a free board of 5.0 m. If k = 6×10^{-4} cm/sec, number of flow channels = 4, number of equipotential drops = 10, calculate the discharge per metre run of the dam.

(ESE: 2010)

Question 9: A 1.5 m thick layer of soil is subjected to an upward seepage head of 1.95 m. What depth of coarse sand will be required above this soil to provide a factor of safety of 1.5 against piping? Coarse sand and soil both have specific gravity 2.65 and porosity 30%.

(ESE: 2010)

Question 10: A masonry dam is founded on previous sand having porosity equal to 45% and specific gravity of sand particles is 2.65. For a desired factor of safety of 3 against sand boiling, determine the maximum permissible upward gradient.

Question 11: A unit volume of a mass of saturated soil is subjected to horizontal seepage. The saturated unit weight is 22 kN/m^3 and the hydraulic gradient is 0.3. Determine the resultant body force on the soil.

Question 12: An earth dam is built on an impervious foundation with a horizontal filter at the base near the toe. The permeability of the soil in the horizontal and vertical directions is 3×10^{-2} mm/s and 1×10^{-2} mm/s respectively. The full reservoir level is 30 m above the filter. A flow net constructed for the transformed section of the dam, consists of 4 flow channels and 16 head drops. Estimate the seepage loss per metre length of the dam.

CHAPTER 6

Consolidation of Soil

Question 6.1: The time taken by a clay layer to achieve 90% consolidation is 15 years. Determine the time required to achieve 90% consolidation, if the layer were twice as thick, three times as permeable and $a_{v2} = 4\, a_{v1}$.

Solution: Since the degree of consolidation U is same for both the cases, $T_V = f(U)$ will also be same for both the cases. Also,

$$T_V = c_V \frac{t}{d^2}$$

$$T_V = \frac{k(1+e_0)t}{a_v \gamma_w d^2} = \text{constant} \tag{1}$$

Let the subscript 1 be used for case 1 and subscript 2 be used for case 2, thus: $t_1 = 15$ years, and $t_2 = $ unknown; $d_2 = 2d_1$; $k_2 = 3k_1$; $a_{v2} = 4\, a_{v1}$. The rest of the variables of equation (1) being constant in both the cases. Since, $T_{V1} = T_{V2}$

$$\frac{k_1 t_1}{a_{v1} d_1^2} = \frac{k_2 t_2}{a_{v2} d_2^2}$$

$$\frac{k_1 t_1}{a_{v1} d_1^2} = \frac{3k_1 t_2}{4 a_{v1} 4 d_2^2}$$

$$t_2 = \frac{16}{3} t_1$$

$$t_2 = \frac{16 \times 15}{3}$$

$$t_2 = 80 \text{ years.}$$

Hence the sample in case 2 will take 80 years to achieve a consolidation (U) of 90%.

Question 6.2: A saturated clay layer of 5 metre thickness takes 1.5 years 50% primary consolidation when drained on both sides. Its coefficient of volume change m_v is 1.5×10^{-3} m²/kN. Determine the coefficient of compressibility (in m²/year) and the coefficient of permeability (in m/year). Assume $\gamma_w = 10$ kN/m³.

Solution: Given: $t = 1.5$ years; $U = 0.5$; $m_v = 1.5 \times 10^{-3}$ m²/kN

$$d = H/2 = 5/2 = 2.5\text{m, for drainage on both sides.}$$

For $U \le 0.6$ or (60%), $T_v = \frac{\pi}{4} U^2 = \frac{\pi}{4}(0.5)^2 = 0.1963$

$$c_v = d^2\frac{Tv}{t} = (2.5)^2 \times \frac{0.1963}{1.5} = 0.818 \text{ m}^2/\text{year}$$

$$k = c_v\, m_v \gamma_w = 0.818 \times 1.5 \times 10^{-3} \times 10 = 0.0123 \text{ m/year.}$$

Question 6.3: In the laboratory test on a clay sample of thickness 25 mm drained at top only, 50% consolidation occurred in 11 minutes. Find the time required for the corresponding clay layer in the field 2 m thick and drained at top and bottom to undergo 70% consolidation. Assume $T_{50} = 0.197$ and $T_{70} = 0.405$.

Solution: Laboratory consolidation:

$H = 25\text{mm} = 0.025\text{m}$

Drainage path for one way consolidation is equal to H.

$$\Rightarrow d = H = 0.025\text{m}$$

Consolidation, $U = 50\% = 0.5$

Time required, $t = 11$ minutes

Coefficient of consolidation is given by

$$c_v = T_{50}\frac{d^2}{t}$$

$$c_v = 0.197 \times \frac{(0.025)^2}{11}$$

$$c_v = 1.1193 \times 10^{-5} \text{ m}^2/\text{minute}$$

(i) Field consolidation

$H = 2\text{m}$

Drainage path for double way consolidation is half of H.

$$\Rightarrow d = H/2 = 1 \text{ m}$$

Time required for consolidation is given as

$$t = T_{70}\frac{d^2}{c_v}$$

$$t = 0.405 \times \frac{(1)^2}{1.1193 \times 10^{-5}}$$

$t = 36183$ min.

$t = 25.127$ days.

Question 6.4: A saturated clay stratum draining both at the top and bottom undergoes 50% consolidation in 16 years under an applied load. If an additional drainage layer were present at the middle of the clay stratum, determine the time (in years) for 50% consolidation.

Solution: Time factor,

$T_v = (C_v \times t) / d^2$ and t is directly proportional to d^2.

d = H for single drainage and d = H/2 for double drainage.

Case 1:

$t_1 = 16$ years, d = H/2

Case 2:

$t_2 = ?$

d = (H/2) / 2 = H/4

Since t is directly proportional to d^2

$$t_1/t_2 = (H/2)^2 / (H/4)^2$$
$$t_2 = 16 \times (H^2/4^{2)}) / (H^2/2^{2)}) = 4 \text{ years.}$$

Question 6.5: A stratum of clay with an average liquid limit of 45% is 6m thick. Its surface is located at a depth of 8m below the ground surface. The natural water content of clay is 40% and its specific gravity is 2.7. Between ground surface and clay, the subsoil consists of fine sand. The water is located at a depth of 4m below the ground surface.

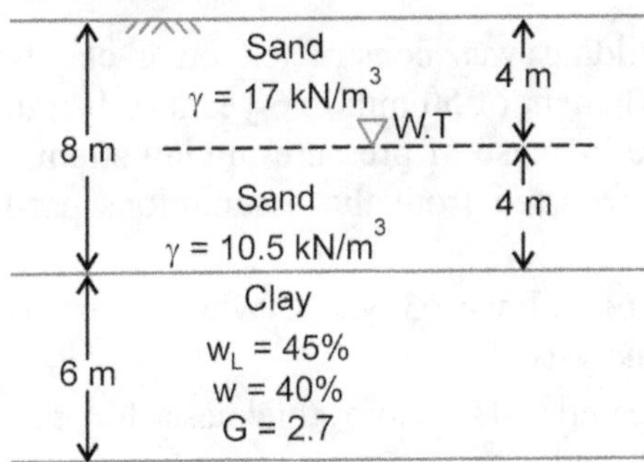

Figure 6.1

The average submerged unit weight of sand is 10.5 kN/m³ and the unit weight of sand above water table is 17 kN/m³. The weight of building that will be constructed on the sand above clay increases the overburden pressure on the clay by 40 kN/m². Estimate the settlement of the building.

Solution: Pressure on the top of clay due to overburden, $\sigma' = 17 \times 4 + 10.5 \times 4$ = 110 kN/m²

Increase in pressure due to construction of building $\Delta\sigma' = 40$ kN/m²

$$C_c = 0.009(w_L - 10) = 0.009(45 - 10) = 0.315$$
$$e_0 = w_{sat}G = 0.4 \times 2.7 = 1.08, H = 6 \text{ m}$$

Settlement

$$S = \frac{C_c\,H}{1+e_0}\log_{10}\frac{\sigma' + \Delta\sigma'}{\sigma'}$$

$$= [(0.315 \times 6) \times \log_{10}(150/110)]/2.08$$

$$= 0.1224\text{m} = 12.24\text{cm}$$

Question 6.6: A 4 m thick layer of normally consolidated clay has an average void ratio of 1.30. Its compression index is 0.6 and coefficient of consolidation is 1 m^2/year. If the increase in vertical pressure due to foundation load on the clay layer is equal to the existing effective overburden pressure, determine the change in the thickness of the clay layer in mm.

Solution: Change in thickness,

$$\Delta H = \frac{C_c\,H}{1+e_0}\log_{10}\frac{\sigma' + \Delta\sigma'}{\sigma'}$$

$C_C = 0.6$, $e_0 = 1.30$, $H_0 = 4$ m and $\Delta \sigma' = \sigma'$

$$\Delta H = \frac{0.6 \times 4}{1+1.3}\,\log_{10}(2/1)$$

$$= 0.314$$

$$= 314 \text{ mm}$$

Question 6.7: A building was constructed on a clay stratum. Preliminary analysis indicated settlement of 50 mm in 6 years and an ultimate settlement of 250 mm. The average increase of pressure in clay stratum was 24 kN/m^2.The following variations occurred from the assumptions used in the preliminary analysis.

   (i)   The loading period was 3 years, which was not considered in the preliminary analysis.

   (ii) Borings indicated 20% more thickness for the clay stratum than originally assumed.

   (iii) During construction, the water table got lowered permanently by 1m.

Estimate:

   1. ultimate settlement

   2. settlement at the end of the loading period

   3. settlement at 2 years after completion of building.

Solution:

(a) Effect of lowering water table

   When water table is lowered, the effective stress gets increased by the amount $(\gamma - \gamma')\,\Delta h$, where γ is the bulk density after lowering water table and Δh is the depth of by which the water table is lowered. Let us assume that $\gamma \approx \gamma_{sat}$, in absence of any other data available.

Hence, $\Delta\sigma' = (\gamma_{sat} - \gamma') \Delta h = \gamma_w \Delta h = 9.81 \times 1 = 9.81$ kN/m²

Now as per preliminary assumptions, $\rho_{f1} = 250$ mm; $\Delta\sigma' = 24$ kN/m² and $(H_0)_1 = H_0$ (say).

For final/actual conditions, $\rho_{f2} = ?$ $\Delta\sigma' = 24 + 9.81 = 33.81$ kN/m² and $(H_0)_2 = 1.2\ H_0$

Now, in general, $\rho_f = m_v . \Delta\sigma' H_0$, or $\rho_{f1} = 250 = m_v\ (24)\ H_0$

or $\qquad m_v\ H_0 = \dfrac{250}{24} = 10.42$ mm/kN/m²

Also, $\quad \rho_{f2} = m_v\ (33.81)\ (1.2\ H_0) = m_v\ H_0\ (33.81 \times 1.2) = 10.42\ (33.81 \times 1.2) = 422.63$ mm

Hence ultimate settlement under actual site conditions will be 422.63 mm instead of 250 mm.

(b) Settlement at the end of construction period

The settlement is assumed to start from mid-time of construction period. Hence settlement at the end of construction period will be that occurring in $3/2 = 1.5$ years.

Now, at $\quad t_1 = 6$ years, $U_1 = 50/250 = 0.2$

Hence at $\quad t_2 = 1.5$ years, $U_2 = U_1\sqrt{\dfrac{t_2}{t_1}} = 0.2\sqrt{1.5/6} = 0.1$

$$\rho_2 = U_2\,\rho_f = 0.1 \times 422.63 = 42.26 \text{ mm}$$

(c) Settlement 2 years after construction period

Here $\qquad t_3 = 2 + 1.5 = 3.5$ years

$\therefore \qquad\qquad U_3 = U_1\sqrt{\dfrac{t_3}{t_1}} = 0.2\sqrt{3.5/6} = 0.1527$

$\therefore \qquad\qquad \rho_3 = U_3\,\rho_f = 0.1527 \times 422.63 = 64.56$ mm

Question 6.8: A clay layer, 8 metre thick, is subjected to a pressure of 70 kN/m². If the layer has a double drainage and undergoes 50% consolidation ($T_v = 0.196$) in one year, determine the coefficient of consolidation. If the coefficient of permeability is 0.40 m/year, determine the settlement in one year. Take $\gamma_w = 9.81$ kN/m³.

Solution: H = 8 m

$\qquad\qquad$ d = H/2 = 8/2 = 4 m.

U = 50%

$\qquad\qquad (T_v)_{50} = 0.196$

t = 1 year

$$\therefore \qquad c_v = T_v \, d^2/t$$

$$= 0.196 \, (4)^2/1$$

$$= 3.136 \text{ m}^2/\text{year}$$

$$m_v = k/c_v \gamma_w$$

$$= 0.040/3.136 \times 9.81$$

$$= 1.3 \times 10^{-3} \text{ m}^2/\text{kN}$$

$$\therefore \qquad \rho_f = m_v \Delta\sigma \, H_0$$

$$= 1.3 \times 10^{-3} \times 70 \times 8$$

$$= 0.728 \text{ m}$$

This is the total settlement.

$\therefore$ Settlement in 1 year = 50% of total settlement = 0.5 x 0.728 = 0.364 m = 364 mm.

Question 6.9: The void ratio of a soil is 0.55 at an effective normal stress of 140 kPa. The compression index of the soil is 0.25. In order to reduce the void ratio to 0.4, determine the increase in the magnitude of effective normal stress in kPa.

Solution: Given, initial void ratio, $e_0 = 0.55$

Initial effective normal stress, $\sigma_1' = 140$ kPa

Compression index, $C_c = 0.25$

Final void ratio, $e_f = 0.4$

Increase in effective normal stress,

$$\Delta\sigma' = ?$$

We know that

$$C_c = [\Delta e/\log_{10}\frac{\sigma' + \Delta\sigma'}{\sigma'}]$$

$$0.25 = [0.15/\log_{10}\frac{140 + \Delta\sigma'}{140}]$$

$$\Delta\sigma' = 417.35 \text{ kPa}$$

Question 6.10: In a consolidation test, the void ratio of the specimen which was 1.068 under the effective pressure of 214 kN/m^2; changed to 0.994 when the pressure was increased to 429 kN/m^2. Calculate the coefficient of compressibility, compression index and the coefficient of volume compressibility. Find the settlement of foundation resting on the above type of clay, if thickness of layer is 8 m and the increase in pressure is 10 kN/m^2.

Solution: Given

$e_0 = 1.068$; $e = 0.994$; $\sigma_0' = 214$ kN/m^2; $\sigma' = 429$ kN/m^2.

Now, coefficient of compressibility, $a_v = \dfrac{\Delta e}{\Delta \sigma'} = \dfrac{e_0 - e}{\sigma' - \sigma_0'}$

$$a_v = \frac{1.068 - 0.994}{429 - 214} = 3.442 \times 10^{-4} \ \text{m}^2/\text{kN}$$

Compression Index $C_c = \dfrac{\Delta e}{\log_{10}\frac{\sigma'}{\sigma_0'}}$

$$C_c = \frac{1.068 - 0.994}{\log_{10}\frac{429}{214}} = 0.245$$

Coefficient of volume compressibility $m_v = \dfrac{a_v}{1 + e_0}$

$$m_v = \frac{3.442 \times 10^{-4}}{1 + 1.068} = 1.664 \times 10^{-4} \ \text{m}^2/\text{kN}$$

Settlement of foundation $\Delta H = \dfrac{C_c H_0}{1 + e_0} \log_{10}\dfrac{\sigma_0' + \Delta\sigma}{\sigma_0'}$

$$\Delta H = \frac{0.245 \times 8}{1 + 1.068} \log_{10}\frac{214 + 10}{214} = 0.0188 \ \text{m} = 18.8 \ \text{mm}.$$

Question 6.11: The settlement analysis for a clay layer draining from top and bottom shows a settlement of 2.5cm in 4 years and an ultimate settlement of 10cm. However, detailed subsurface investigation reveals that there is no drainage at the bottom. Determine the ultimate settlement in this condition.

Solution: The ultimate settlement does not depend upon drainage conditions. Hence, the ultimate settlement remains as 10cm.

Question 6.12: The initial and final void ratios of clay sample in a consolidation test are 1 and 0.5, respectively. If initial thickness of sample is 2.4cm, then determine its final thickness.

Solution: $H_f/H_i = (1 + e) / (1 + e_o)$

$H_f = (1 + 0.5) / (1 + 1) \times 2.4$

$= 1.5 \times 1.2 = 1.8$cm.

Question 6.13: A clay sample originally 26mm thick at a void ratio of 1.22, was subjected to a compressive load. After the clay sample was completely consolidated, its thickness was measured to be 24mm. What is the final void ratio?

Solution: $\Delta H/H_i = \Delta e / (1 + e_o)$

$(26 - 24) / 26 = \Delta e / (1 + 1.22)$

$$\Delta e = 0.171$$
$$e_1 - e_2 = \Delta e$$
$$e_2 = 1.22 - 0.171$$
$$e_2 = 1.05.$$

Question 6.14: A 1m thick layer of saturated clay drained at both faces settles by 10cm in one year. If a thin layer of pervious soil is introduced in the middle of this layer, then what will be the period during which settlement of 10cm will be completed?

Solution: For a particular soil, t is directly proportional to d^2

$$t_1/t_2 = (d_1/d_2)^2 \text{ or } 1/t_2 = (1/0.5)^2 \text{ or } t_2 = 0.25 \text{ year}$$

Question 6.15: Draw e – log p curve using the following data:

Pressure σ' (kg/cm^2)	0.25	0.50	1	2	4	8	2	0.5	0.05
Dial change (mm)	0.2945	0.410	0.460	1.275	2.620	3.830	3.700	3.353	3.19

Initial height of specimen = 26 mm; Final height of specimen = 23.1045 mm

Final void ratio = 0.57515605

Solution: The calculations are given in the table below:

$$\Delta h = (D_{c1} - D_{c2}) = 0.2945 - 0.41 = -0.1155 \text{ mm.}$$
$$h = h_i - \Delta h$$
$$= 26 - 0.1155$$
$$= 25.8845 \text{ mm.}$$

Change in void ratio, $\Delta e = -(\Delta h)(1 + e_f)/h_f$
$$= (1 + 0.57515605)/23.1045 \times \Delta h$$
$$= 0.0681753 \times \Delta h$$
$$= 0.0681753 \times (-0.1155)$$
$$= -0.00787425$$

Void ratio, $e = (e_f - \Delta e) = 0.57515605 - 00.01111257 = 0.56404348$

Pressure σ'		Dial change D_c(mm)	$\Delta h = (D_{c1} - D_{c2})$ (mm)	Specimen height at the end of each loading $(h_i - \Delta h)$ (mm)	Change in void ratio $\Delta e = (\Delta h)(1 + e_f)/h_f = 0.0681753 \times \Delta h$	Void ratio $e_f = (e_i + \Delta e)$
(kg/cm²)	kPa					
0.25	25	0.2945		26		0.77255763
			−0.1155		−0.00787425	
0.5	50	0.41		25.8845		0.76468338
			−0.05		−0.00340877	
1	100	0.46		25.8345		0.76127462
			−0.815		−0.05556287	
2	200	1.275		25.0195		0.70571175
			−1.345		−0.09169578	
4	400	2.62		23.6745		0.61401597
			−1.21		−0.08249211	
8	800	3.83		22.4645		0.53152386
			0.13		0.00886279	
2	200	3.7		22.5945		0.54038665
			0.347		0.02365683	
0.5	50	3.353		22.9415		0.56404348
			0.163		0.01111257	
0.05	5	3.19		23.1045		0.57515605↑

e – log p curve is plotted as given below. The values of C_c and C_s are calculated as under:

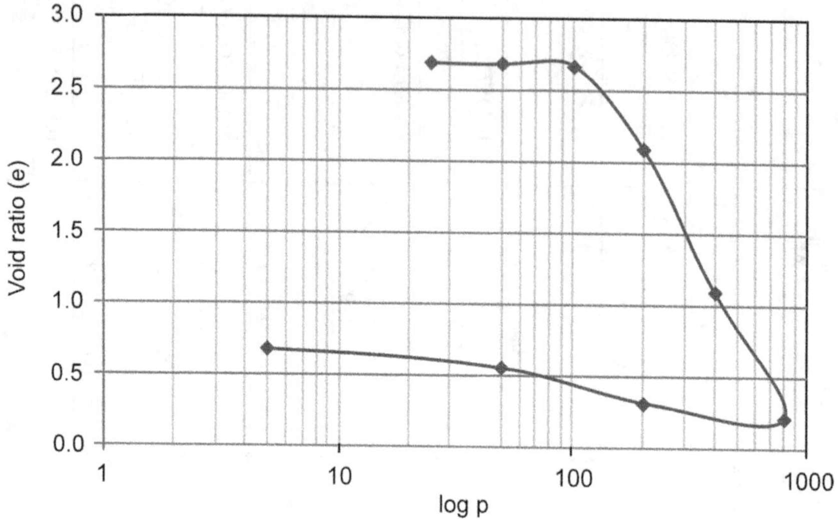

Figure 6.2: e – log p curve.

$$C_c = \frac{\Delta e}{\log_{10}\frac{\sigma'}{\sigma_0'}}$$

$$= \frac{2.66932203 - 1.00881084}{\log_{10}\frac{8}{1}} = 1.84$$

$$C_s = \frac{\Delta e}{\log_{10}\frac{\sigma'}{\sigma_0'}}$$

$$= \frac{1.00881084 - 0.57515605}{\log_{10}\frac{0.05}{8}} = 0.197.$$

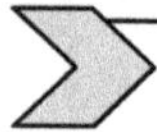

Questions

Question 1: In a consolidation test, the void ratio of the specimen which was 2.068 under the effective pressure of 316 kN/m^2 changed to 0.999 when pressure was increased to 529 kN/m^2. Calculate the coefficient of compressibility, compression index and the coefficient of volume compressibility. Find the settlement of foundation resting on above type of clay, if thickness of layer is 7 m and the increase in pressure is 14kN/m^2.

Question 2: In the laboratory test on a clay sample of thickness 35 mm drained at top only, 40% consolidation occurred in 10 minutes. Find the time required for the corresponding clay layer in the field 3 m thick and drained at top and bottom to undergo 60% consolidation. Assume $T_{50} = 0.197$ and $T_{60} = 0.405$.

Question 3: A footing 2m square, rests on a soft clay soil with its base at a depth of 1.5m below the ground surface. The clay stratum is 3.5m thick and is underlainby a firm sand stratum. The clay soil has LL = 30%, G = 2.7; water content at saturation = 40%; cohesion 0.5 kg/cm^2 (ø = 0). It is known that the clay stratum is normally consolidated. Compute the settlement that would result if the load intensity equal to safe bearing capacity of soil were allowed to act on footing. Natural water table is quite close to the ground surface. For the given conditions, bearing capacity factor (N_c) is obtained as 6.9.Take factor of safety as 3. Assume load spread of 2 (vertical) : 1 (horizontal).

(ESE: 1995)

Question 4: In a consolidation test, the void ratio of the specimen which was 1.09under the effective pressure of 200kN/m^2, changed to 0.99 when the pressure was increase to 400 kN/m^2. Calculate the coefficient of compressibility, compression index and coefficient of volume compressibility. Find the settlement of foundation resting on above type of clay if thickness of layer is 8 m and the increase in pressure is 10 kN/m^2.

Question 5: Under a certain loading, a layer of clay is expected to undergo full settlement of 18cm. Also, it is expected to settle by 5 cm in the period of first 2 months of loading. Find the time required for the clay layer to settle by 10cm.

(ESE: 2004)

Question 6: A soil of specific gravity 2.67 has a moisture content of 20%.When fully saturated 2.0 cm thick sample of this soil tested in a consolidometer shows a compression of 0.050 cm when the load is increased from 45 kN/m^2 to 90 kN/m^2. Compute the compression index of the soil.

(ESE: 2008)

Question 7: How many days would be required by a clay stratum 5 m thick, draining at both ends with coefficient of consolidation = 50 × 10^{-4} cm^2/sec to attain 50% of its ultimate settlement?

(ESE: 2008)

Question 8: In a consolidation test done in laboratory a sample of 20mm thick consolidated 50% in 15 minutes under double drainage. How much time a 5.0 m thick layer of same soil will require to attain 50% and 30% consolidation? If the soil layer has a rock below, how much time it will take to consolidate 50% and 30%?

(ESE: 2010)

Question 9: A saturated clay stratum draining both at the top and bottom undergoes 50 percent consolidation in 16 years under an applied load. If an additional drainage layer were present at the middle of the clay stratum, determine the time required for 50 percent consolidation.

Question 10: In an oedometer test, a specimen of saturated clay 19 mm thick reaches 50% consolidation in 20 minutes. How long it would take a layer of this clay 5 mm to reach the same degree of consolidation under the same stress and drainage conditions? How long would it take for the clay layer to reach 30% consolidation?

Question 11: The time required for a clay layer to achieve 85% consolidation is 10 years. If the layer was half as thick, 10 times more permeable and 4 times more compressible then determine the time that would be required to achieve the same degree of consolidation.

Question 12: A double drainage clay layer, 6 m thick, settles by 30 mm in three years under the influence of a certain loads. Its final consolidation settlement has been estimated to be 120 mm. If a thin layer of sand having negligible thickness is introduced at a depth of 1.5 m below the top surface, determine the final consolidation settlement of clay layer.

Question 13: An 8 m thick saturated clay layer lies above a permeable dense sand layer. The clay settles by 40 mm in 2 years when subjected to a widespread load of 50 kN/m^2 at its surface.

  (i) What will be the ultimate settlement of the clay?

  (ii) Calculate the compression index, C_c of the clay.

  Assume that the water table is below the dense sand layer and will not influence the settlement. Properties of clay: Average bulk unit weight, γ_{bulk}= 18 kN/m^3, coefficient of consolidation c_v= 1 m^2/year and specific gravity, G = 2.65. Unit weight of water is 9.81 kN/m^3

Question 14: A clay soil specimen 25 mm thick was tested in the laboratory allowing drainage both sides. The specimen showed a decrease in void ratio from 0.92 to 0.78 when the pressure was increased from 60kN/m^2 to 120kN/m^2. Calculate the coefficient of compressibility, coefficient of volume change and coefficient of consolidation. If time for 50% consolidation was 5 minutes; determine the soil permeability in cm/sec and calculate final thickness of the specimen at the end of the test.

Question 15: The base of a 20 m x 50 m raft rests on ground surface and applies a stress of 100 kPa to the subsoil. The ground strata consist of dense sand up to 4 m depth followed by a clay layer of 2 m thickness having compression index of 0.40 and initial void ratio of 1.0. The clay is followed by more dense sand. Find the maximum settlement that will occur due to consolidation of clay. Water table is at the ground surface and the total unit weight of sand is 20 kN/m^3 and of clay is 18 kN/m^2. Calculations should be based on two equal divisions. Use 2:1 stress distribution.

Question 16: In a consolidated undrained test with pore pressure measurement on normally consolidated clay, a sample consolidated under a stress of 200kPa failed at an additional axial stress of 150kPa. The pore pressure at failure was 75kPa. Determine analytically the shear stress parameters in terms of effective stress.

Question 17: At a vertical stress of 200 kPa, the void ratio of a saturated soil sample tested in an oedometer is 1·52 and lies on the normal consolidation line. An increment of vertical stress of 150 kPa in the second stage compresses the sample to a void ratio of 1.43.

(a) Determine the compression index C_c of the soil.

(b) Sample was unloaded to a vertical stress of 200 kPa and the void ratio increased to 1.45. Determine the slope of the recompression index C_r.

(c) What is the over consolidation ratio of the soil at second stage?

(d) If the soil was reloaded to a vertical stress of 500 kPa, what void ratio would be obtained?

CHAPTER 7

Compaction

Question 7.1: A sample of sand has volume of 1000 ml in its natural state. Its minimum volume when compacted is 750 ml. When gently poured into a measuring cylinder. Its possible maximum volume is 1320 ml. What is the relative density?

Solution: $V = 1000$ ml

$V_{min} = 750$ ml

$V_{max} = 1320$ ml

$e = $ void ratio $= V_v/V_s$

Hence, $V_s = $ constant for all conditions only V_v changes.

Let, $V_s = x$

$e = V_v/V_s = (V - V_s)/ V_s = (V/V_s) - 1$

For natural state, $e = (1000/x) - 1$

For maximum compacted state,

$e_{max} = (1320/x) - 1$

For minimum compacted state

$e_{min} = (750/x) - 1$

Relative density $= [(e_{max} - e)/ (e_{max} - e_{min})] \times 100$

$100 \times [\{(1320/x) - 1\} - \{(1000/x) - 1\}] / [\{(1320/x) - 1\} - \{(750/x) - 1\}]$

$= 56.14\%$

Question 7.2: A cohesive soil yields a maximum dry density of 1.8 g/cm^3 at an OMC of 16% during a standard Proctor test. If the value of G is 2.65, what is the degree of saturation? What is the maximum dry density it can further be compacted to?

Solution: Given

Maximum dry density $\rho = 1.8$g/cm^3; OMC $= w = 0.16$; and $G = 2.65$

$e = (G\rho_w/\rho) - 1$

$= \{(2.65 \times 1)/1.8\} - 1 = 0.4722$

Degree of saturation $= S = w\,G/e$

$$= (0.16 \times 2.56) / 0.4722$$

$$= 0.8979 = 89.79\%$$

Hence the degree of saturation is 89.79%

Now, $\rho = G\rho_w / (1+wG)$

$$= 2.65 \times 1 / (1+0.16 \times 2.65)$$

$$= 1.861 \text{ g/cm}^3$$

Hence the maximum dry density is 1.861 g/cm^3

Question 7.3: A soil deposit has a void ratio of 0.9.If the void ratio is reduced by 0.6 due to compaction, determine the percentage volume loss due to compaction.

Solution: Given $e_1 = 0.9$, $e_2 = 0.6$.

Let us use symbol 1 for initial condition and symbol 2 for final compacted condition.

Initially,

$$V_{v1}/V_s = e_1 = 0.9; \text{ hence } V_{v1} = 0.9V_s$$

Hence, $V_1 = V_{v1} + V_s = 0.9V_s + V_s = 1.9\,V_s$ .....(1)

Finally,

$$V_{v2}/V_s = e_2 = 0.6; \text{ hence } V_{v2} = 0.6V_s$$

Hence, $V_2 = V_{v2} + V_s = 0.6V_s + V_s = 1.6\,V_s$ .....(2)

Percentage loss in volume $= [(1.9\,V_s - 1.6V_s) / 1.9\,V_s] \times 100$

$$= 15.79\,\%.$$

Question 7.4: A soil has maximum dry unit weight of 18 kN/m^3 at a moisture content of 16% during a standard Proctor test. What is the degree of saturation of soil if its specific gravity is 2.65?

Solution: $\gamma_{d\,max} = 18$ kN/m^3

$w = 16\%$, $G = 2.65$

$\gamma_{d\,max} = G\gamma_w/(1 + e)$

$18 = 2.65 \times 9.81/ (1 + e)$

$e = 0.44$

$S_e = wG$

$S \times 0.44 = 16 \times 2.65$

$S = 95.5\%$

Question 7.5: From the following observations during a standard Proctor test, determine the maximum dry density and optimum moisture content:

Mass of Proctor mould = 2069 g

Volume of Proctor mould = 1000 cm^3

Observation No.	1	2	3	4	5
Mass of mould + compacted soil (g)	3593	3677	3782	3845	3812
Mass of container (g) = W_0	15.72	16.28	16.10	15.13	15.86
Mass of container + wet W_1 soil (g)	47.59	41.23	40.42	38.46	45.48
Mass of container + dry W_2 soil (g)	44.27	38.32	37.29	35.03	40.42

Solution:

Volume of Proctor mould V = 1000 cm^3

Observation No.	1	2	3	4	5
Mass of Proctormould + compacted soil (g)	3593	3677	3782	3845	3812
Mass of Proctor mould (g)	2069	2069	2069	2069	2069
Mass of compacted soil (g) M_C	1524	1608	1713	1776	1743
Wet density $\rho = \frac{M_c}{V}$ (g/cm^3)	1.524	1.608	1.713	1.776	1.743
Mass of container + wet soil M_1 (g)	47.59	41.23	40.42	38.46	45.48
Mass of container + dry soil M_2 (g)	44.27	38.32	37.29	35.03	40.42
Mass of water (g) = $(M_1 - M_2)$	3.32	2.91	3.13	3.43	5.06
Mass of container (g) = M_0	15.72	16.28	16.10	15.13	15.86
Mass of dry soil (g) $M_d = (M_2 - M_0)$	28.55	22.04	21.19	19.90	24.56
Water content $w = \frac{M_w}{M_d} \times 100$	11.63	13.20	14.77	17.24	20.60
Dry density $\rho_d = \frac{\gamma_t}{1+w}$	1.365	1.420	1.492	1.515	1.445

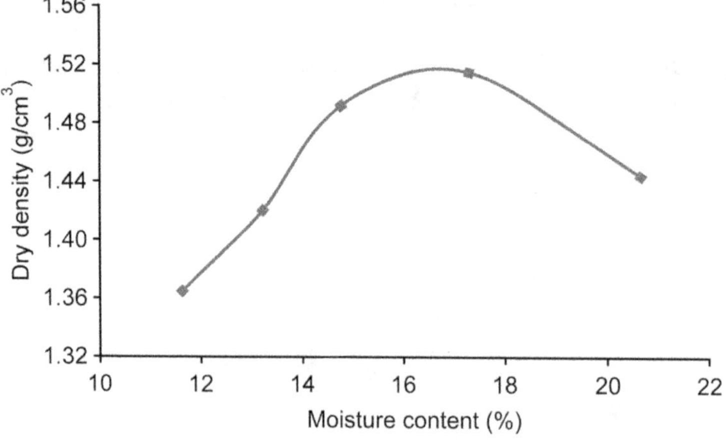

Figure 7.1: Dry density – moisture content curve.

From graph

Maximum dry density $= 1.518$ g/cm^3

Optimum moisture content $= 16.6\%$.

Question 7.6: The data on the variation of maximum dry density and optimum moisture content of soft clayey soil upon addition of different percentages of waste foundry sand is given below:

Waste foundry sand (%)	0	10	20	30	40
Optimum moisture content (%)	16.5	17.2	17.7	18.0	18.3
Maximum dry density ρ_{dmax}(g/cm^3)	1.71	1.72	1.78	1.79	1.80

Plot the graph showing the variation of maximum dry density and optimum moisture content with varying percentages of waste foundry sand.

Solution:

The optimum moisture content increased from 16.5% to 18.3% when waste foundry sand was added to clayey soil in proportions varying from 10% to 40%.

The maximum dry density enhanced from 1.71 to 1.80 g/cm^3 with waste foundry sand content increasing from 10% to 40%.

The variation of maximum dry density and optimum moisture content with varying percentages of waste foundry sand is shown in figure 7.2.

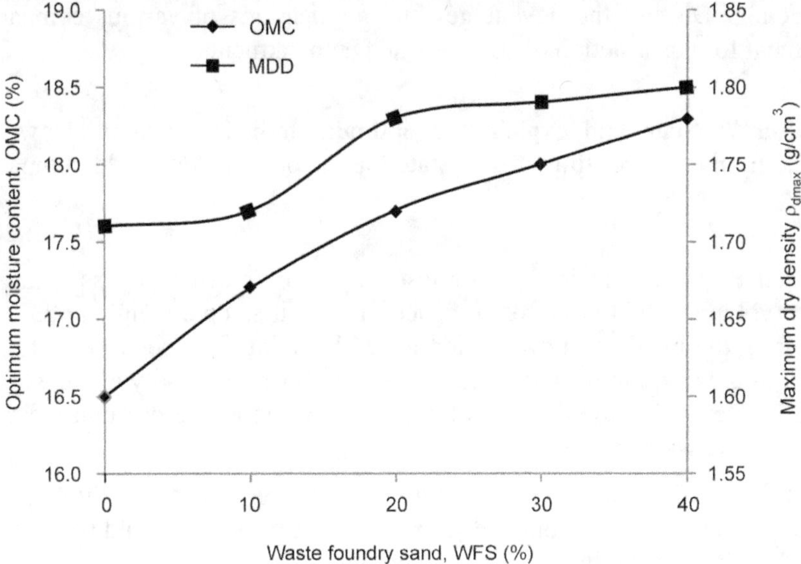

Figure 7.2: Maximum dry density and optimum moisture content with waste foundry sand.

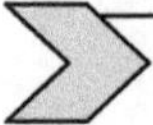

Questions

Question 1: Work out the theoretical maximum dry density for a soil sample having specific gravity of 2.7 and OMC = 16%. Also explain the difference in OMC values in case of Procter test and modified Proctor test for cohesive soil and granular soil.

(ESE: 2001)

Question 2: In a Proctor compaction test, the soil specimen of one of the observations had a bulk density of 19 kN/m^3 at a moisture content of 15%. Determine degree of saturation of the specimen, if G = 2.7 additional moisture content required for saturating the soil specimen.

(ESE: 2004)

Question 3: A soil has maximum dry unit weight of 18.5 kN/m^3 at moisture content of 19% during the standard Proctor test. What is the degree of saturation of soil if its specific gravity is 2.65?

Question 4: For the compaction of which type of soils: sheep foot rollers, and vibratory rollers are the most effective? Explain giving reason.

(ESE: 2006)

Question 5: State and explain different factors influencing the compaction of a soil in the field.

(ESE: 2005)

Question 6: Explain the objects of compaction and effect of inadequate compaction. Discuss the advantages and applications of various compacting equipment for construction of subgrade and embankments.

(ESE: 2007)

Question 7: State and explain the optimum moisture content. How is it affected by the compacting effort? State the factors affecting field compaction of soil.

(ESE: 2009)

Question 8: In a standard proctor test, 1.8 kg of moist soil was filling the mould (volume = 944 cc) after compaction. A soil sample weighing 23 g was taken from the mould and oven-dried for 24 hours at a temperature of 110 °C. Weight of the dry sample was found to be 20 g. Specific gravity of soil solids is G = 2.7. Determine the theoretical maximum value of the dry unit weight of the soil at that water content.

Question 9: Indian Standard (IS) Light Compaction test is usually conducted in 3 layers in a 1000 cm^3 mould. How many blows per layer would be necessary if the above test is conducted in a 2250 cm^3 mould?

CHAPTER 8

Shear Strength of Soil

Question 8.1: An unconfined compression test yielded strength of 0.1 N/mm^2. If the failure plane is inclined at 50° to the horizontal, what are the values of the shear strength parameters?

Solution: For unconfined compression test,

$$\sigma_3 = 0$$

$$q = \sigma_1 = 2c \tan[45° + \phi/2]$$

$$\alpha_c = 50°$$

$$50° = 45° + \phi/2$$

$$\phi = 10°$$

$$q = 0.1 \, \text{N/mm}^2$$

$$0.1 = 2c \times \tan(50°)$$

$$c = 0.1/2(\tan 50°)$$

$$c = 0.0419 \, \text{N/mm}^2$$

Shear strength parameters c and ϕ are 0.0419 N/mm^2 and 10° respectively.

Question 8.2: A triaxial shear test conducted on the sand specimen at a confining pressure 100 kN/m^2 under drained condition resulted in a deviator stress $(\sigma_1' - \sigma_3')$ at failure equal to 100 kN/m^2. Determine the angle of shearing resistance of the soil.

Solution: $\sigma_1' = \sigma_3' + \sigma_d'$

$$= 100 + 100 = 200 \, \text{kN/m}^2$$

For sand c = 0

$$\sigma_1' = \sigma_3' \tan 2\alpha + 2c \tan \alpha$$

Where $\alpha = 45° + \phi/2$

$$200 = 100 \tan 2\alpha$$

$$\alpha = 54.73° = 45° + \phi/2$$

$$\phi = 19.47°$$

Question 8.3: In an unconsolidated undrained triaxial test, it is observed that an increase in cell pressure from 150 kPa to 250 kPa leads to a pore pressure increase of 80 kPa. It is further observed that, an increase of 50 kPa in deviatoric stress results in an increase of 25 kPa in the pore pressure. Determine the value of Skempton's pore pressure parameter B.

Solution: Given

Increase in cell pressure $\Delta\sigma_3 = 250 - 150$

$$= 100 \text{ kPa.}$$

Increase in pore pressure $\Delta u = 80$ kPa.

Skempton's pore pressure parameter B $= \Delta u/\Delta\sigma_3$

$$= 80/100 = 0.8$$

The value of Skemptons's pore pressure parameter B $= 0.8$

Question 8.4: A 20 m thick clay layer is sandwiched between a silty sand layer and a gravelly sand layer. The layer experiences 30 mm settlement in 2 years. If the coefficient of consolidation of the layer is 0.003 cm²/s, determine the time required for 50 mm settlement of the deposit.

Solution: Given:

$$T_v = (\pi/4) \times (U/100)^2 \text{, for } U \leq 60\%; \text{ and}$$
$$= \{1.781 - 0.993\log_{10}(100 - U), \text{ for } U > 60\%$$

Where T_v is the time factor and U is the degree of consolidation in percentage.

$$T_v = C_v\, t/d^2 = (0.003{\times}2{\times}86400{\times}365)/\,[(20/2) \times 100]^2$$
$$= 0.189$$

Also, $T_v = (\pi/4) \times (U/100)^2 = 0.189$

$U = 49\% \leq 60\%$

Consolidation settlement $= 30/0.49 = 61.22$mm

Degree of consolidation for 50mm settlement,

$$U = 50/61.22 = 0.817 = 81.7\%$$

Now $T_v = \{1.781 - 0.993 \log_{10}(100 - U),$

Or $0.608 = C_v\, t/d^2$

Or $t = (0.608 \times 10^2)/\,(0.003 \times 10^{-4})$

$$= 202666667 \text{ s}$$
$$= 6.43 \text{ years}$$

Additional time required $= 6.43 - 2 = 4.43$ years.

Question 8.5: In an unconfined compression test on saturated clay, the undrained shear strength was found to be 6 t/m^2. If the sample of the same soil is tested in an undrained condition in triaxial compression at a cell pressure of 20 t/m^2, determine the major principal stress at failure.

Solution: In an UCS test on clay,

Undrained shear strength $(c_u) = 6$ t/m^2

For same soil in triaxial test,

$$\sigma_3 = 20 \text{ t/m}^2$$

$$\sigma_1 = \sigma_3 \tan^2(45° + \phi/2) + 2c \tan(45° + \phi/2)$$

For clay, $\phi = 0$

$$\sigma_1 = \sigma_3 + 2c_u$$

$$= 20 + 2 \times 6$$

$$= 32 \text{ t/m}^2.$$

Question 8.6: A laboratory vane shear test apparatus is used to determine the shear strength of a clay sample and only one end of the vane takes part in shearing the soil. If T = applied torque, H = height of vane and D = diameter of vane, then give the formula for shear strength of clay.

Solution: For vane shear test with one end in shearing,

$$\tau = T / \pi[D^2(H/2 + D/12)]$$

When both ends take part in shearing,

$$\tau = T / \pi[D^2(H/2 + D/6)]$$

Question 8.7: A soil fails under an axial vertical stress of 100 kN/m^2 in unconfined compression test. The failure plane makes an angle of 50° with the horizontal. Determine the shear parameters c and ϕ.

Solution: The angle of failure with horizontal

$$\alpha = 45° + \phi/2$$

$$\phi/2 = (50° - 45°)$$

$$\phi = 10°$$

$$\sigma_1 = 100 \text{ kN/m}^2$$

$$\sigma_3 = 0$$

$$\sigma_1 = \sigma_3 \tan^2(45° + \phi/2) + 2c \tan(45° + \phi/2)$$

$$\sigma_1 = 2c \tan(45° + \phi/2)$$

$$100 = 2c \tan 50°$$

$$c = 100 / (2c \tan 50°)$$

$$= 41.9 \text{ kN/m}^2$$

Question 8.8: A clay soil specimen when tested in unconfined condition gave an unconfined compressive strength of 100 kN/m^2. A specimen of the same clay with the same initial condition is subjected to UU triaxial test under a cell pressure of 100 kN/m^2. Determine the axial stress.

Solution: Unconfined compressive strength

$$q_u = 100 \ kN/m^2$$

For UU test,

$$\sigma_1 - \sigma_3 = q_u$$
$$\sigma_1 = 100 + 100 = 200 \ kN/m^2$$

Question 8.9: If s is the shear strength, c and ϕ are shear strength parameters; and 'σ_n' is the normal stress at failure, then give the expression for c based upon coulomb's equation.

Solution: From Coulomb's equation

$$s = c + \sigma_n \tan \phi$$
$$c = s - \sigma_n \tan \phi$$

Question 8.10: A dry sand specimen is tested in a triaxial test. The cell pressure is 50 kPa and the deviator stress at failure is 100 kPa. Determine the angle of internal friction for sand specimen.

Solution: For sand,

Cohesion c = 0

$$\sigma_1 = \sigma_3 \tan^2(45° + \phi/2)$$
$$\sigma_1 = \sigma_3 + \sigma_d$$
$$\sigma_3 = 50 \ kPa \ and \ \sigma_1 = 100 \ kPa$$

Now, $$\sigma_1 = 50 + 100 = 150 \ kPa$$
$$\tan^2(45° + \phi/2) = \sigma_1/\sigma_3 = 150/50 = 3$$
$$(45° + \phi/2) = 60°$$
$$\phi = 30°$$

Question 8.11: If the unconfined compressive strength of 4 kg/cm^2 in the natural state of clay reduces by four times in the remolded state, then determine its sensitivity.

Solution: Sensitivity = $(UCS)_{natural}$ / $(UCS)_{remoulded}$
$$= 4/(4/4)$$
$$= 4$$

Question 8.12: A CD triaxial test was conducted on a regular soil. At failure σ_1 / σ_3 was 3.0. The effective minor principal stress at failure was 75 kPa. Determine the principal stress difference at failure.

Solution: $\sigma_3' = 75$ kPa; $\sigma_1' = 3\sigma_3'$

Principal stress difference

$$= \sigma_1' - \sigma_3'$$

$$= 3\sigma_3' - \sigma_3' = 2\sigma_3'$$

$$= 2 \times 75$$

$$= 150 \text{ kPa}$$

Question 8.13: In an unconfined compression test on stiff clay, if the failure plane made an angle of 52° to the horizontal, what would be the angle of shearing resistance?

Solution: If the failure plane makes an angle α with the horizontal, then

$$\alpha = 45° + \phi/2$$

Where ϕ is the angle of shearing resistance

$$52° = 45° + \phi/2$$

$$\phi = 14°$$

Question 8.14: A dry sand specimen is tested in a triaxial test. The cell pressure is 75 kPa and the deviator stress at failure is 150 kPa. Determine the angle of internal friction for sand.

Solution: $\sigma_c = \sigma_3 = 75$ kPa

$$\sigma_1 = \sigma_3 + \sigma_d$$

$$= 75 + 150$$

$$= 225 \text{ kPa}$$

For sand c = 0,

$$\sigma_1 = \sigma_3 \tan^2(45° + \phi/2)$$

$$225 = 75 \tan^2(45° + \phi/2)$$

$$\tan^2(45° + \phi/2) = 3$$

$$\tan(45° + \phi/2) = \sqrt{3}$$

$$(45° + \phi/2) = 60°$$

$$\phi = 30°.$$

Question 8.15: The following data is obtained from consolidated undrained triaxial test conducted on a soil sample. Plot the curve between volumetric strain and axial strain.

Length of sample (m) = 0.076; Diameter of sample (m) = 0.038. Least count = 0.00001.

Change in volume = 0.0000007 m^3.

Deformation dial gauge reading	Volume change reading	Deformation dial gauge reading	Volume change reading
0	32.1	280	33.2
20	32.1	300	33.3
40	32.1	320	33.6
60	32.1	340	33.8
80	32.1	360	33.9
100	32.1	380	34
120	32.2	400	34.1
140	32.3	420	34.2
160	32.3	440	34.4
180	32.5	460	34.5
200	32.6	480	34.6
220	32.9	500	34.7
240	32.9	520	35
260	33	540	35

Solution:

Length of sample L = 0.076 m. Diameter of sample d = 0.038 m. Least count = 0.00001.

Area of sample = $\pi \times (0.038)^2/4 = 0.001134$ m^2.

Volume $V_0 = \pi L \times (0.038)^2/4 = 8.6184 \times 10^{-5}$ m^3.

Change in volume = $0.0000007 = 0.07 \times 10^{-5}$ m^3.

Corrected volume $V_c = 8.6184 \times 10^{-5} - 0.07 \times 10^{-5} = 8.5484 \times 10^{-5}$ m^3.

Axial strain $\varepsilon = \Delta L/L \times 100$

= Dial reading × least count × 100/L

= 20 × 0.00001 × 100/0.076

= 0.2631.

Volumetric strain $\varepsilon_v = \Delta V/V \times 100$

= (Volume change reading − initial volume change reading) × $10^{-6} \times 100/V$

= (32.2 − 32.1) × $10^{-6} \times 100 /8.6184 \times 10^{-5}$

= 0.11698.

The axial strain and volumetric strain values are given in the table below.

Deformat ion dial gauge reading	Volume change reading	Axial strain $\varepsilon\Delta L/L$ (%)	Volumetric strain ε_v $\Delta V/V$ (%)	Deformation dial gauge reading	Volume change reading	Axial strain $\varepsilon\ \Delta L/L$ (%)	Volumetric strain $\varepsilon_v\Delta V/V$ (%)
0	32.1	0	0	280	33.2	3.6842	1.2867905
20	32.1	0.2631	0	300	33.3	3.9473	1.4037715
40	32.1	0.5263	0	320	33.6	4.2105	1.7547143
60	32.1	0.7894	0	340	33.8	4.4736	1.9886762
80	32.1	1.0526	0	360	33.9	4.7368	2.1056572
100	32.1	1.3157	0	380	34	5	2.2226382
120	32.2	1.5789	0.116981	400	34.1	5.2631	2.3396191
140	32.3	1.8421	0.2339619	420	34.2	5.5263	2.4566001
160	32.3	2.1052	0.2339619	440	34.4	5.7894	2.690562
180	32.5	2.3684	0.4679238	460	34.5	6.0526	2.8075429
200	32.6	2.6315	0.5849048	480	34.6	6.3157	2.9245239
220	32.9	2.8947	0.9358476	500	34.7	6.5789	3.0415048
240	32.9	3.1578	0.9358476	520	35	6.8421	3.3924477
260	33	3.4210	1.0528286	540	35	7.1052	3.3924477

The results are plotted to obtain the curve between volumetric strain and axial strain.

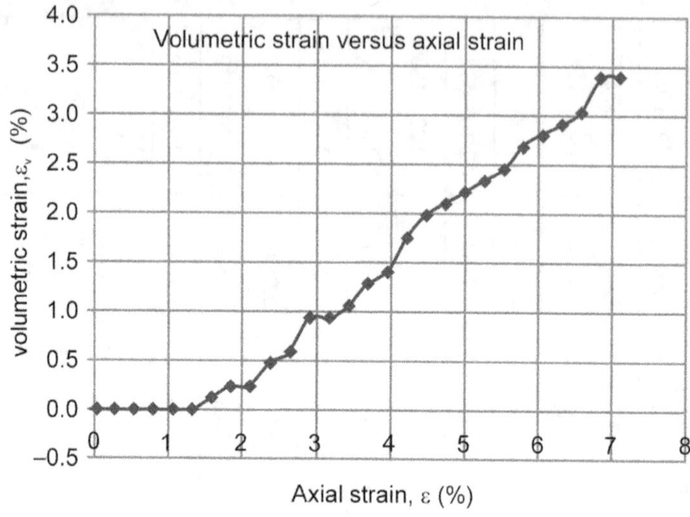

Figure 8.1: Volumetric strain versus axial strain for consolidated undrained triaxial test.

Question 8.16: Plot the following data to obtain $q - p$ plot for cohesionless soil:

S. No.	p (kPa)	q (kPa)
1	57.07	37.02
2	85.08	55.65
3	128.5	84.1

Solution: The $q - p$ data is plotted as shown in the figure 8.2 below.

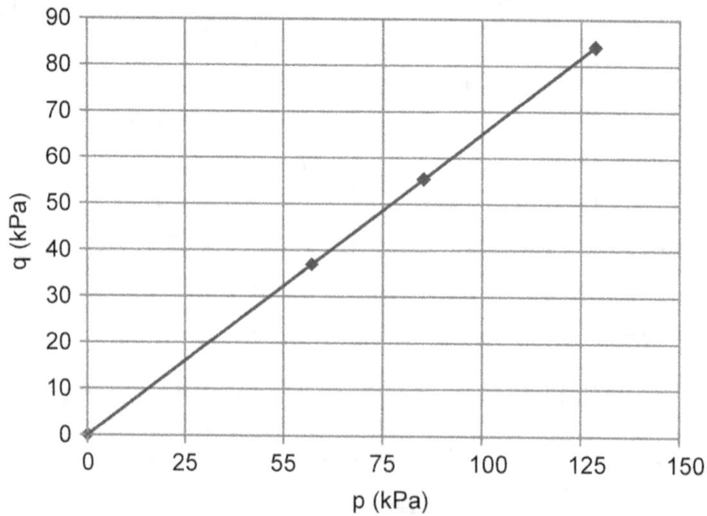

Figure 8.2: $q - p$ plot for consolidated undrained triaxial test.

From the curve, $\psi' = 34.21°$.

$$\varphi' = \sin^{-1}(\tan \psi') = 42.83°.$$

Question 8.17: The following data is obtained from consolidated undrained triaxial test conducted on a soil sample:

Length of sample (m) = 0.076; Diameter of sample (m) = 0.038. Least count = 0.00001.

Change in volume = 0.0000007 m^3. Proving ring constant = 0.0030053

Least count of dial gauge = 0.00001. Cell pressure σ_3 = 29.43 kPa.

Plot the stress-strain curve for the soil sample.

Deformation dial gauge reading	Proving ring reading	Deformation dial gauge reading	Proving ring reading
0	0	280	46.5
20	1	300	47.5
40	17	320	48
60	26	340	49
80	31	360	49.5
100	34.5	380	50.5
120	37.25	400	50.5
140	39.5	420	50.5
160	41.5	440	51
180	42.5	460	51
200	44	480	51
220	46	500	51
240	46	520	51
260	46	540	51

Solution: Length of sample L = 0.076 m.

Diameter of sample d = 0.038 m.

Least count of dial gauge = 0.00001.

Proving ring constant PR = 0.0030053

Area of sample A = $\pi \times (0.038)^2/4 = 0.001134$ m^2.

Axial strain $\varepsilon = \Delta L/L \times 100$

$\qquad = $ Dial reading $\times$ least count $\times$ 100/L

$\qquad = 20 \times 0.00001 \times 100/0.076 = 0.2631$.

Deviator stress$\sigma_3 = P \times PR/A_c$

$\qquad = 17 \times 0.0030053/0.00114$

$\qquad = 44.815877$ kPa.

Deformation dial gauge reading	Proving ring reading P	Axial strain $\varepsilon\Delta L/L$	Corrected area $A_c = A/(1-\varepsilon)$ (m^2)	Deviator stress σ_3 $P \times PR/A_c$ (kPa)	Deformation dial gauge reading	Proving ring reading P	Axial strain $\varepsilon\Delta L/L$	Corrected area $A_c = A/(1-\varepsilon)$ (m^2)	Deviator stress σ_3 $P \times PR/A_c$ (kPa)
0	0	0	0.001134	0	280	46.5	3.6842	0.001193	106.90604
20	1	0.2631	0.001137	2.6432022	300	47.5	3.9473	0.001197	108.77448
40	17	0.5263	0.00114	44.815877	320	48	4.2105	0.001205	107.48624
60	26	0.7894	0.001143	68.360602	340	49	4.4736	0.001211	108.54789
80	31	1.0526	0.001146	81.290673	360	49.5	4.7368	0.001215	109.31015
100	34.5	1.3157	0.001149	90.228044	380	50.5	5	0.00122	111.2953
120	37.25	1.5789	0.001154	94.866413	400	50.5	5.2631	0.001225	110.80898
140	39.5	1.8421	0.001158	98.775929	420	50.5	5.5263	0.00123	110.32378
160	41.5	2.1052	0.001161	104.14498	440	51	5.7894	0.001236	110.32066
180	42.5	2.3684	0.001167	103.63831	460	51	6.0526	0.001241	109.85945
200	44	2.6315	0.001171	106.33768	480	51	6.3157	0.001246	109.3972
220	46	2.8947	0.001179	107.76711	500	51	6.5789	0.001251	108.93415
240	46	3.1578	0.001182	108.24215	520	51	6.8421	0.001259	107.18799
260	46	3.4210	0.001187	107.45383	540	51	7.1052	0.001262	107.38442

The results are tabulated in the table above.

The stress-strain curve for the soil sample is plotted below.

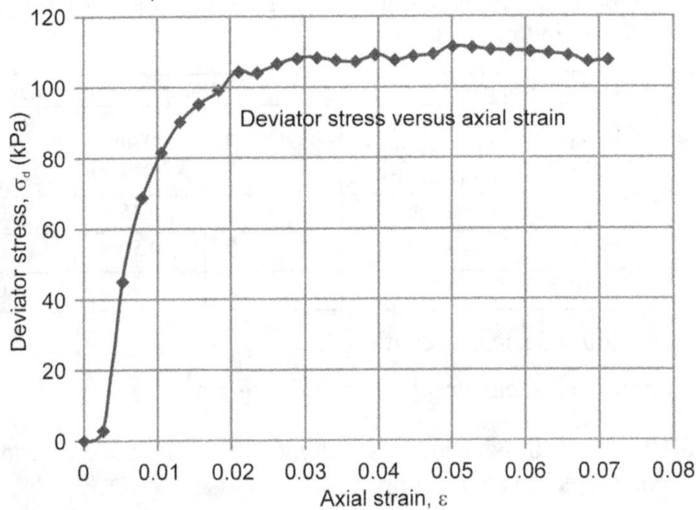

Figure 8.3: Deviator stress - axial strain curve for consolidated undrained triaxial test.

Question 8.18: The vane shear test data for undisturbed and remoulded soil is given below:

Undisturbed specimen		Remoulded specimen	
Specimen No.	**Torque T (kg-cm)**	**Specimen No.**	**Torque T (kg-cm)**
1	1.05	1	0.45
2	1.15	2	0.50
3	1.10	3	0.45

Diameter of blade = 1.2 cm, and height of the blade = 2.4 cm. Determine the undisturbed and remoulded shear strength of soil.

Solution: The shear strength S of soil can be determined by using the expression:

$$S = \frac{T}{\frac{\pi D^2}{2}\left(H + \frac{D}{3}\right)}$$

$$= 0.1578916\ T$$

$$= \frac{3}{19}\ T$$

Where

S = shear strength (kg/cm^2)

T = torque in kg-cm

D = diameter of blade = 1.2 cm, and

H = height of the blade = 2.4 cm.

Using the given data, the undisturbed and remoulded shear strength of soil is calculated as given in table below:

Undisturbed specimen			Remoulded specimen		
Specimen No.	Torque T (kg-cm)	Undisturbed shear strength S (kg/cm^2)	Specimen No.	Torque T (kg-cm)	Remoulded shear strength S (kg/cm^2)
1	1.05	0.166	1	0.45	0.071
2	1.15	0.182	2	0.50	0.079
3	1.10	0.174	3	0.45	0.071

Average undisturbed shear strength = 0.174 kg/cm^2.

Average remoulded shear strength = 0.0737 kg/cm^2.

Question 8.19: The following data is obtained from unconfined compressive strength test conducted on a soil sample:

Length of sample = 76 mm;

Diameter of sample = 38 mm.

Proving ring constant: 1 division = 15 N.

Least count of dial gauge = 0.01 mm.

Compression dial reading	0	10	20	30	40	50	60	70	80	90	100	110	120
Proving ring reading	0	5.2	8.8	13.2	17.4	22.4	26.4	32.6	34.6	38.9	42.1	48.6	51.3

Plot the stress-strain curve for the soil sample.

Solution: Given data:

Length of sample = 76 mm;

Diameter of sample = 38 mm.

Proving ring constant: 1 division = 15 N.

Least count of dial gauge = 0.01 mm.

Area of cross section of sample $A_0 = \frac{\pi}{4} D_0^2$

$$= 11.34 \text{ cm}^2$$
$$= 0.001134 \text{ m}^2.$$

The deformation, load, corrected area strain and stress values are calculated as given in the table. The unconfined compressive strength-strain curve for the soil sample is plotted below.

Compression dial reading	Proving ring reading	Deformation (mm)	Strain ε $= (\Delta L/L) \times 100$ (%)	Area (m^2) $A_c = \dfrac{A_0}{1-\epsilon}$	Axial load P (N)	Compressive stress P/A_c (kPa)
0	0	0	0	0.001134	0	0
10	5.2	0.1	0.0132	0.001135	78	68.69256
20	8.8	0.2	0.0262	0.001137	132	116.0958
30	13.2	0.3	0.0394	0.001138	198	173.914
40	17.4	0.4	0.0526	0.001140	261	228.9474
50	22.4	0.5	0.0658	0.001142	336	294.347
60	26.4	0.6	0.0789	0.001143	396	346.4495
70	32.6	0.7	0.0921	0.001145	489	427.2452
80	34.6	0.8	0.1053	0.001146	519	452.8544
90	38.9	0.9	0.1184	0.001148	583.5	508.4569
100	42.1	1	0.1316	0.001149	631.5	549.551
110	48.6	1.1	0.1447	0.001151	729	633.5526
120	51.3	1.2	0.1579	0.001152	769.5	**667.8571**

Deformation ΔL = compression dial reading × dial gauge least count

$$= 10 \times 0.01$$

$$= 0.1 \text{ mm.}$$

Axial load P = proving ring reading × proving ring constant

$$= 5.2 \times 15$$

$$= 78 \text{ N.}$$

Unconfined compressive strength = 667.8571 kPa.

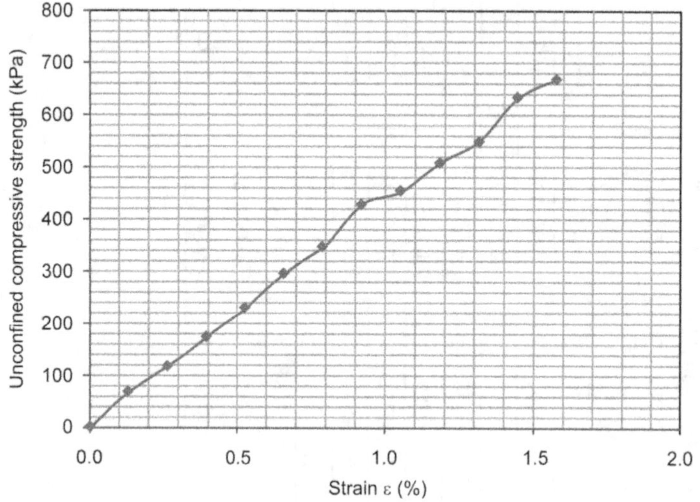

Figure 8.4: Unconfined compressive stress-strain curve.

Question 8.20: The following data is obtained from direct sheartest conducted on a sand sample:

Proving ring constant = 3.0053 N;

Least count of dial gauge = 0.01 mm

Normal stress = 0.25 kg/cm^2;

Relative density = 0.3 or 30%

Shear dial reading	0	80	160	240	320	400	480	560	640	720	800	880	960	1040	1120	1200
Proving ring reading	0	9	16	19	24	24	24	21	20	20	20	20	20	20	20	20

Plot the shear stress – shear displacement curve for the sand sample.

Solution: Given data:

Proving ring constant = 3.0053 N;

Least count of dial gauge = 0.01 mm

The shear displacement, corrected area and shear stress values are calculated as given in the table below:

Shear displacement = shear dial reading × dial gauge least count

$$= 80 \times 0.01$$

$$= 0.8 \text{ mm.}$$

Corrected area $A_c = 6 \times (6 - SD/10)$

$$= 6 \times (6 - (0.8/10))$$

$$= 35.52 \text{ cm}^2.$$

Shear stress = proving ring reading × proving ring constant/A_c

$$= 9 \times 3.0053 \times 10000/(35.52 \times 1000)$$

$$= 7.61 \text{ kPa.}$$

Shear dial reading	Shear displacement SD (mm)	Corrected area (cm²) $A_c = 6 \times (6 - SD/10)$	Proving ring reading	Shear stress (kpa)
0	0	36	0	0
80	0.8	35.52	9	7.61
160	1.6	35.04	16	13.72
240	2.4	34.56	19	16.52
320	3.2	34.08	24	21.16
400	4	33.60	24	21.47
480	**4.8**	**33.12**	**24**	**21.78**
560	5.6	32.64	21	19.34
640	6.4	32.16	20	18.69
720	7.2	31.68	20	18.97
800	8	31.20	20	19.26
880	8.8	30.72	20	19.57
960	9.6	30.24	20	19.88
1040	10.4	29.76	20	20.20
1120	11.2	29.28	20	20.53
1200	12	28.80	20	20.87

The shear stress – shear displacement curve for the sand sample is plotted below.

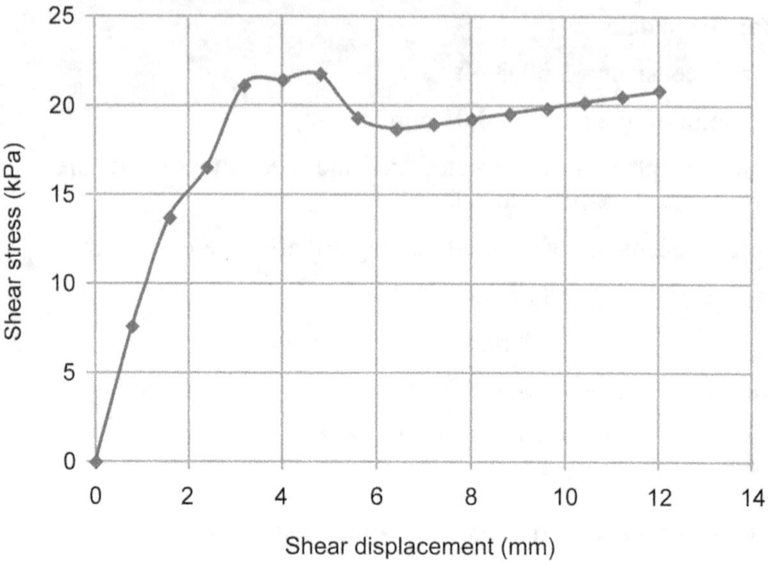

Figure 8.5: Shear stress – shear displacement curve for sand sample.

Questions

Question 1: A consolidated undrained test was conducted on soil sample with cell pressure $\sigma_c = 100$ kN/m^2. The deviator stress at failure $(\sigma_1 - \sigma_3)$ was observed to be 60 kN/m^2. The soil is known to have a cohesion $c' = 0$ and the angle of shearing resistance $\phi = 30°$ and an undrained cohesion $c_u = 0$ and angle of shearing resistance, $\phi_u = 13.3°$. What was the pore water pressure at failure?

Question 2: Two identical specimens of a backfill soil, when tested in a triaxial compression test, gave following stresses (in kg/cm^2) at failure:

Sample No.	All round pressure	Axial stress	Pore water pressure
1	1.0	2.2	0.6
2	2.0	3.6	1.1

1. Find the axial stress at which the third identical specimen of the same soil will fail if all- round pressure is 2.8 kg/cm^2.

2. Find the total and effective shear strength parameters of the soil.

(ESE: 2004)

Question 3: What is the significance of pore water pressure in a triaxial shear test? Explain the pore water pressure parameters as given by Skempton.

(ESE: 2004)

Question 4: Samples of dry sand are to be tested in triaxial and direct shear tests. In the triaxial test, specimen fails when the major and minor principal stresses are 960 kN/m^2 and 260 kN/m^2 respectively. What shear strength would be expected in the direct shear test when the normal stress is 230 kN/m^2?

(ESE: 2002)

Question 5: A dry sand specimen is tested in a tri-axial test. The cell pressure is 70 kPa and the deviator stress at failure is 140 kPa. Determine the angle of internal friction for sand.

Question 6: For a triaxial shear test conducted on a sand specimen at a confining pressure of 150 kN/m^2 under drained condition resulted in a deviator stress ($\sigma_1 - \sigma_3$) at failure of 150 kN/m^2. Determine the angle of shearing resistance of the soil.

Question 7: In an unconfined compression test on saturated clay, the undrained shear strength was found to be 8 t/m^2. If the sample of the same soil is tested in an undrained condition in triaxial compression at a cell pressure of 10t/m^2, then determine the major principal stress at failure.

Question 8: The undrained cohesion of a remoulded clay soil is 10 kN/m^2. If the sensitivity of the clay is 20, determine the corresponding remoulded compressive strength.

Question 9: In an unconsolidated undrained triaxial test, it is observed that an increase in cell pressurefrom 150 kPa to 250 kPa leads to a pore pressure increase of 80 kPa. It is further observedthat, an increase of 50 kPa in deviatoric stress results in an increase of 25 kPa in the porepressure. Determine the value of Skempton's pore pressure parameter B.

Question 10: A tri-axial shear test conducted on a sand specimen at a confining pressure of100kN/m^2 under drained conditions resulted in a deviator stress ($\sigma_1 - \sigma_3$) at failureof100kN/m^2. Determine the angle of shearing resistance of the soil.

Question 11: A sample of saturated cohesion less soil tested in a drained triaxial compression test showed an angle of internal friction of 30°. Determine the deviatroic stress at failure for the sample at a confining pressure of 200 kPa.

Question 12: A clay soil sample is tested in triaxial apparatus in consolidated-drained conditions at a cell pressure of 100 kN/m^2. What will be the pore water pressure at a deviator stress of 40kN/m^2?

Question 13: A direct shear test was conducted on a cohesion-less soil (c=0) specimen under a normal stress of 200kN/m^2. The specimen failed at a shear stress of 100kN/m^2. Determine the angle of internal friction of the soil.

Question 14: A CU triaxial compression test was performed on saturated sand at a cell pressure of 100 kPA. The ultimate deviator stress was 350 kPa and the pore pressure at the speak stress was 40 kPs (suction). Estimate the total and effective stress shear strength parameters.

Question 15: In a triaxial test carried out on a cohesionless soil sample with a cell pressure of 20 kPa, the observed value of applied stress at the point of failure was 40 kPa. Determine the angle of internal friction of the soil.

Question 16: If the effective stress strength parameters of a soil are $c' = 10$ kPa and $\phi' = 30°$. Determine the shear strength on a plane within the saturated soil mass at a point where the total normal stress is 300 kPa and pore water pressure is 150 kPa.

Question 17: A conventional consolidated drained (CD) triaxial test was conducted on saturated clean sand sample by using the following steps:

1. Set cell pressure to 250 kPa and allow the sample to consolidate with its drainage valve open at 100 kPa back pressure.

2. Shear the sample without any change in the drainage condition. The sample failed when the deviator stress reached 300 kPa.

Use analytical solution to determine:
   (i)   The slope of failure envelope in degrees. Assume c' to be zero.
   (ii) Slope of the failure plane in degrees.
   (iii) Shear stress and normal stress on the failure plane (in kPa).
   (iv) The maximum shear stress at failure (in kPa).

Question 18: A sample of normally consolidated clay was subjected to a consolidated undrained triaxial compression test that was carried out until the specimen failed at a deviator stress of 50 kN/m^2. The pore water pressure at failure was recorded to be 20 kN/m^2 and confining pressure of 50 kN/m^2 was used in the test. Determine the consolidated undrained friction angle and drained friction angle.

Question 19: When an unconfined compression test was conducted on a cylindrical soil sample, it failed under an axial stress of 120 kN/m^2. The failure plane makes an angle of 50° with the horizontal. Determine the cohesion and the angle of internal friction of the soil.

Question 20: Determine the axial stress at failure for a dry dense sand in triaxial loading if $\sigma_3 = 300$ kN/m^2. A previous test had given $\sigma_3 = 150 kN/m^2$; $\sigma_1 = 735 kN/m^2$ at failure.

CHAPTER 9

Stress Distribution in the Soil

Question 9.1: A concentrated load of 50 t acts vertically at a point on the soil surface. If the Boussinesq's equation is applied for computing the stress, then determine the ratio of vertical stresses at depth of 3m and 5m respectively vertically below the point of application of load.

Solution: Vertical stress below the point load,

$$\sigma_z = 0.4775(Q/Z^2)$$
$$\sigma_3/\sigma_5 = (5/3)^2 = 2.77$$

Question 9.2: For vertical concentrated load acting on the surface of a semi-infinite elastic soil mass, determine the vertical normal stress at depth 'z'.

Solution: According to Boussinesq's theory,

$$\sigma_z = 3Q/2\pi z^2[1/1 + (r/z^2)]^{5/2}$$
$$= K_b Q/z^2$$

Question 9.3: In the Newmark's influence chart for stress distribution, there are 10 concentric circles and 50 radial lines. Determine the influence factor of the chart.

Solution: Influence factor $= 1/10 \times 50 = 0.002$

Question 9.4: σ_z is the vertical stress at a depth equal to Z in the soil mass due to a surface point load Q. Determine the vertical stress at a depth equal to 2Z.

Solution: Under the point load, the vertical stress is given by

$$\sigma_z = 0.4775Q/Z^2$$

(From Boussinesq's equation at $r/z = 0$)

$$\therefore \sigma_z = 0.4775Q/4Z^2 = 0.25\ \sigma_z$$

The depth at which $\sigma_z = 0.1$ per unit area is

$Z = \sqrt{4.775} = 2.185$ units under the load.

Thus, isobar extends up to 2.185 units of depth under the load to dissipate 90% of stresses.

Question 9.5: Determine the change in the vertical stress in the soil mass estimated by Boussinesq's equation when Poisson's ratio of soil changes from 0.3 to 0.5.

Solution: Vertical stress at a point due to a point load Q is

$$\therefore \sigma_z = 3Q/2\pi Z^2 \left[1/(1+(x/z)^2)\right]^{5/2}$$

It depends only on r/z and is independent of Poisson's ratio. So, there won't be any change in vertical stress.

Question 9.6: In a Newmark's chart the stress distribution, there are 10 concentric circles and 20 radial lines. Determine the influence factor for the chart.

Solution: Influence factor = 1/ (No. of concentric circles × No. of radial lines)

$$= 1/(10 \times 20) = 0.005$$

Question 9.7: Determine the value of r/z for which Westergaard's formula for vertical stress gives greater value of stress than that by the Boussinesq's formula.

Solution: The Boussinesq's equation gives vertical stress

$$\sigma_z = K_B \times Q/Z^2$$

Where $\quad K_B = 3/2\pi/ \left[1+(r/z)^2\right]^{5/2}$ .....(i)

For $\mu = 0$, the Westergaard's solution is

$$\sigma_z = K_w \times Q/Z^2$$

$$K_w = 1/\pi/ \left[1+ 2(r/z)^2\right]^{3/2}$$ ....(ii)

Equating (i) and (ii), we get,

$$r/z = 1.524$$

For $\quad r/z < 1.524; \ K_w < K_B$

Westergaard's formula gives lesser value of vertical stress. For r/z > 1.524, $K_w > K_B$ i.e., Westergaard's formula gives greater value of stress.

Question 9.8: A point load of 650kN is applied on the surface of a thick layer of clay. Using Boussinesq's elastic analysis, what is approximate value of the estimated vertical stress at a depth 2 m and a radial distance of 1.0 m from the point of application of load?

Solution: According to Boussinesq's theory,

$$\sigma_z = 3Q/2\pi Z^2 \left[1/(1+(r/z)^2)\right]^{5/2}$$

$$= 3 \times 650/2\pi \times 2^2 \left[1+ (1/2)^2\right]^{5/2}$$

$$= 44.4 \ kN/m^2.$$

Question 9.9: A line load of infinite length has an intensity q per unit length. What is the vertical stress σ_z at a depth z below the earth at the center of the load?

Solution: Vertical stress at a point P due to a line load of intensity q/unit length.

$$\sigma_z = 2q/\pi Z[1/(1+(x/Z)^2)]^2$$

Where,

X = distance of point from centerline of load

Z = depth of point below the center of load

Question 9.10: A concentrated load of 50 kN acts on the surface of ground. Determine the increase in vertical stress directly below the load at a depth of 4 m. (Value of influence factor is 0.48).

Solution:

$$\sigma_z = (Q/Z^2)IB = Q/Z^2\,(0.48)$$
$$= 50/4^2\,(0.48) = 1.5 \text{ kN/m}^2$$

Question 9.11: What is the Boussinesq's vertical stress at a point 6 m directly below a concentrated load of 2000 kN applied at the ground surfaces?

Solution:

$$\sigma_z = 0.4775 \times Q/Z^2 = 0.4775 \times 2000 / (6)^2 = 26.5 \text{ kN/m}^2.$$

Question 9.12: The vertical stress at any point at a radial distance r and at depth z as determined by using Boussinesq's influence factor K_B and Westergaard's influence factor K_W would be almost same for what (r/z) ratios?

Solution: For r/z = 1.5

K_B and K_W would be almost same

Questions

Question 1: An elevated structure with a total weight of 10,000 kN is supported on a tower with 4 legs. The legs rest on piers located at the corners of a square 6m on a side. What is the vertical stress increment due to this loading at a point 7 m beneath the centre of the structure? Assume that the load be approximated to a point load acting at the corners of a square of 6 m side.

(ESE: 2012)

Question 2: What is the Boussinesq's vertical stress at a point 8 m directly below a concentrated load of 4000 kN applied at the ground surfaces?

Question 3: A concentrated load of 70 kN acts on the surface of ground. Determine the increase in vertical stress directly below the load at a depth of 6 m. (Value of influence factor is 0.58).

Question 4: A point load of 750 kN is applied on the surface of a thick layer of clay. Using Boussinesq's elastic analysis, what is approximate value of the estimated vertical stress at a depth 2.8 m and a radial distance of 1.4 m from the point of application of load?

Question 5: In a Newmark's chart the stress distribution, there are 15 concentric circles and 25 radial lines. Determine the influence factor for the chart.

Question 6: Determine the change in the vertical stress in the soil mass estimated by Boussinesq's equation when Poisson's ratio of soil changes from 0.35 to 0.5.

CHAPTER 10

Effective Stress

Question 10.1: Steady state seepage is taking place through a soil element at Q, 2 m below the ground surface immediately downstream of the toe of an earthen dam as shown in the figure 10.1. The water level in a piezometer installed at P, 500 mm above Q, is at the ground surface. The water level in piezometer installed at R, 500 mm below Q, is 100 mm above the ground surface. The bulk saturated unit weight of the soil is $10kN/m^2$ and the unit weight of water is $9.81kN/m^2$. Determine the vertical effective stress (in kPa) at Q.

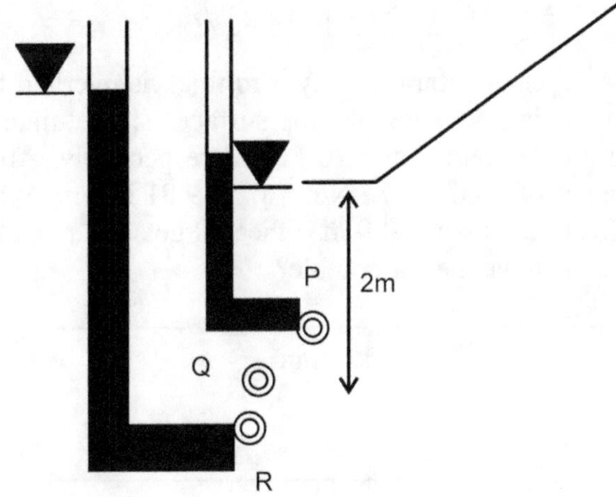

Figure 10.1

Solution: At P, h = 1.5 m

At R, h = 2.6 m

At   Q, h = (1.5 + 2.6) / 2 = 2.05 m

Effective stress at Q, $\sigma' = \sigma - u$

$$= \gamma_{sat}H - \gamma_w H$$

$$= 18 \times 2 - 9.81 \times 2.05$$

$$= 15.89 kN/m^2.$$

Question 10.2: A layer of saturated clay 5 m thick is over lain by sand 4.0 m deep. The water table is 3 m below the top surface. The saturated unit weights of clay and sand are 18 kN/m^3 and 20 kN/m^3 respectively. Above the water table, the unit weight of sand is 17 kN/m^3. Given γ_w= 9.81 kN/m². Determine the effective pressures on a horizontal plane at a depth of 9 m below the ground surface.

Solution: Total pressure at 9 m below the ground surface

$$\sigma = 17 \times 3 + 20 \times 1 + 18 \times 5$$

$$\sigma = 161 \text{ kN/m}^2$$

Pore water pressure at 9 m below the ground surface

$$u = 9.81 \times 1 + 5 \times 9.81$$

$$u = 58.86 \text{ kN/m}^2$$

Effective pressure

$$\bar{\sigma} = \sigma - u$$

$$\bar{\sigma} = 161 - 58.86 \text{ or } \bar{\sigma} = 102.14 \, kN/m^2$$

Question 10.3: A layer of saturated clay 5 m thick is overlaid by sand 4.0 m deep. The water table is 3 m below the top surface. The saturated unit weights of clay and sand are 18 kN/m^3 and 20 kN/m^3 respectively. Above the water table, the unit weight of sand is 17 kN/m^3. γ_w = 9.81 kN/m². What will be the increase in the effective pressure at 9 m if the soil gets saturated by 4 capillary, up to a height of 1 m above the water table?

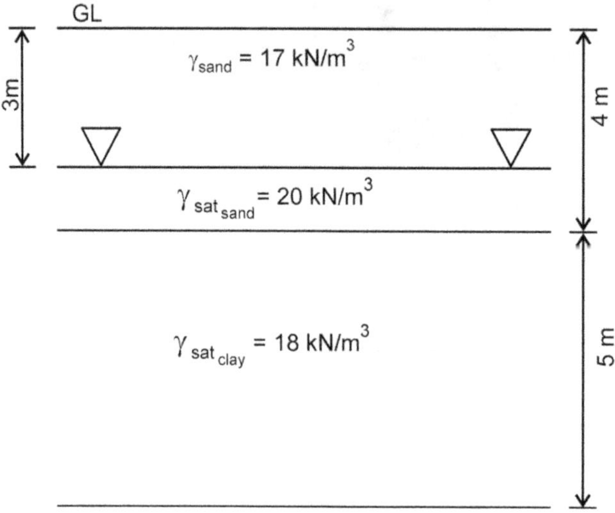

Figure 10.2

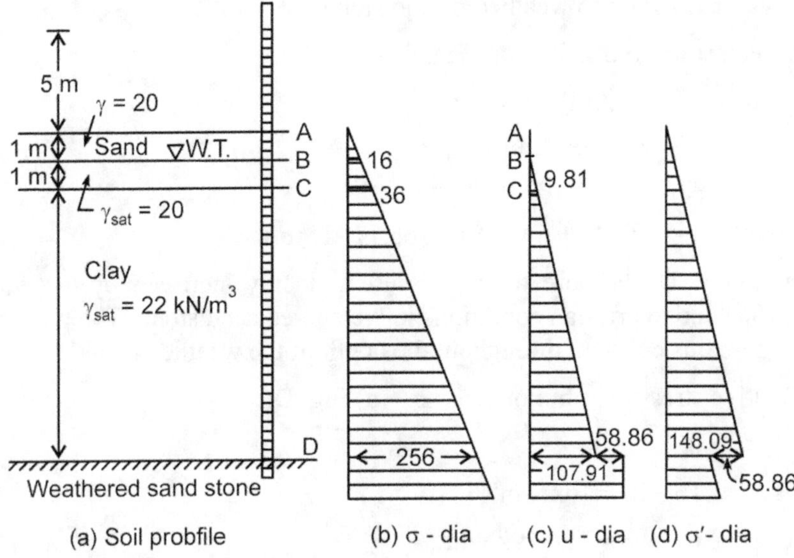

Figure 10.3

Solution: If the soil gets saturated by capillarity, then

$$\sigma' = 2 \times 17 + (1 + 1)(20 - 9.81) + 1 \times 9.81 + 5(18 - 9.81)$$

$$\sigma' = 105.14 \ kN/m^2$$

Increase in pressure $= 105.14 - 102.14 \ = 3 \ kN/m^2$

Therefore,

Increase in pressure $= 3 kN/m^2$

Question 10.4: The soil profile at the bottom of the valley comprises 2 m of sand overlying 10 m of clay. The clay layer itself is resting on highly permeable weathered sand stone. The unit weight of sand above the water table, which is 1 m below the ground level, is 16 kN/m³, and below the water table, it is 20 kN/m³. The saturated unit weight of clay is 22 kN/m³. If the water table in the sand stone layer is under artesian condition, corresponding to a stand pipe level at 5 m above the ground level, plot the variation of (i) total stress, (ii) pore water pressure, and (iii) effective stress, with depth.

Solution: At A, $\sigma = 0$;

At   B, $\sigma = 16 \ \times 1 = 16 kN/m^2$

At   C, $\sigma = 16 + (20 \times 1) = 36 kN/m^2$;

At   D, $\sigma = 36 + 22 \times 10 = 256 kN/m^2$;

Below D, $\sigma = 256 + \gamma \ z$;

where γ and z refer to weathered sand stone.

(ii) Pore Pressure (u) distribution: Fig. c

At A,u = 0;   At B,u = 0;

At C, u = 9.81 × 1 = 9.81 kN/m²;

At D, u = 9.81 × 11 = 107.91 kN/m²

Below D, u = 107.91 + 6 × 9.81= 166.77 kN/m²

Thus below D, the pore water pressure suddenly increases by 6 × 9.81 = 58.56kN/m² due to artesian conditions in weathered sand stone. The pore water pressure remains constant throughout the depth of the weathered sold stone.

(iii) Effective stress distribution ($\sigma' = \sigma - u$) Fig. D

At A,$\sigma' = 0$;

At B $\sigma' = 16 - 0 = 16$kN/m²

At C, $\sigma'=36 - 9.81 = 26.19$kN/m²;

At C, $\sigma'=256 - 107.91 = 148.09$kN/m²;

Below D, $\sigma'=148.09 - 6 \times 9.81=89.23$ kN/m²

Thus, below D, σ' suddenly decreases by amount equal to excess artesian pressure $(6 \times \gamma_w)$.

Question 10.5: The ground conditions at a site are shown in the figure 10.4 below.

Find the saturated unit weight of the sand (kN/m³) and the total stress, pore water pressure and effective stress (kN/m²) at the point P, respectively.

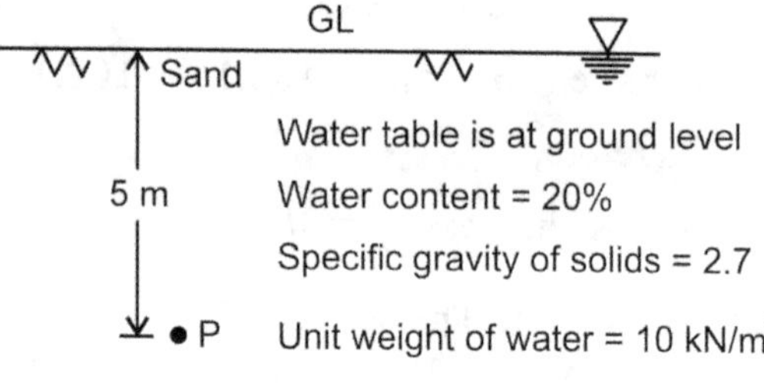

Figure 10.4

Solution:

$$e = \frac{Gw}{S} = \frac{2.7 \times 0.2}{1} = 0.54$$

$$\gamma_{sat} = \frac{G+e}{1+e} \times \gamma_w$$

$$\gamma_{sat} = \frac{2.7+0.54}{1+0.54} \times 10 = 21 \ kN/m^3$$

Total stress, $\sigma = \gamma_{sat}h = 21.0 \times 5 = 105$ kN/m²

Pore water pressure, u $= \gamma_w h = 10 \times 5 = 50$ kN/m²

Effective stress, $\bar{\sigma} = \sigma - u = 105 - 50 = 55$ kN/m².

Question 10.6: A river 5 m deep consists of a sand bed with saturated unit weight of 20 kN/m^3, $\gamma_w = 9.81 kN/m^3$. Determine the effective vertical stress at 5 m from the top of sand bed.

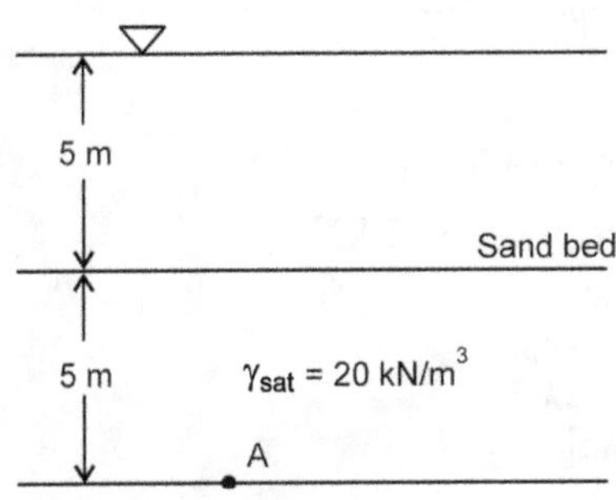

Figure 10.5

Solution: Total stress at A,

$$\sigma_A = \gamma_w z_1 + \gamma_{sat} z_2$$
$$= 9.81 \times 5 + 20 \times 5 = 149.05 \; kN/m^2$$

Neutral stress at A,

$$u_A = \gamma_w(z_1 + z_2) = 9.81 \times 10 = 98.1 \; kN/m^2$$

Effective stress at A,

$$\bar{\sigma} = \sigma - u = 149.05 - 98.1 = 50.95 \; kN/m^2$$

Question 10.7: A sand layer found at sea floor under 20 m water depth is characterised with relative density = 40%, maximum void ratio = 1.0, minimum void ratio = 0.5, and specific gravity of soil solids = 2.67. Assume the specific gravity of sea water to be 1.03 and the unit weight of fresh water to be 9.81 kN/m^3. What would be the effective stress at 30 m depth into the sand layer?

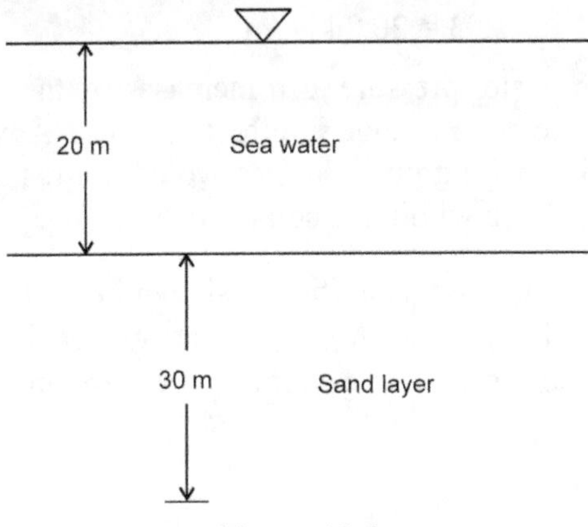

Figure 10.6

Solution: Relative density, I = 40%

Maximum void ratio, $e_{max} = 1.0$

Minimum void ratio, $e_{min} = 0.5$

Specific gravity of soil G = 2.67

Specific gravity of sea water, $G_{sw} = 1.03$

Unit wt. of fresh water, $\gamma_w = 9.81$ kN/m^3

$$I_D = \frac{e_{max}-e}{e_{max}-e_{min}}$$

$$0.4 = \frac{1.0-e}{1.0-0.5}$$

$$e = 0.80$$

$$\gamma_{sat} = \left(\frac{G+e}{1+e}\right)\gamma_w$$

$$\therefore \gamma_{sat} = \left(\frac{2.67+0.8}{1+0.8}\right) \times 9.81$$

$$\therefore \gamma_{sat} = 18.91 \ kN/m^3$$

Effective stress at 30m below sand layer

$$\bar{\sigma} = (\gamma_{sat} - \gamma_w)G_{sw}H = (18.91 - 9.81) \times 1.03 \times 30 = 281 \text{ kPa}$$

Question 10.8: A sand layer found at sea floor under 20 m water depth is characterised with relative density = 40%, maximum void ratio = 1.0, minimum void ratio = 0.5, and specific gravity of soil solids = 2.67. Assume the specific gravity of sea water to be 1.03 and the unit weight of fresh water to be 9.81 kN /m^3. What would be the effective stress at 30 m depth into the sand layer?

Solution: A rise of sea water by 2 m would result in an increase in total stress by $2\gamma_w$

$$= 2 \times 9.81 \times 1.03 = 20.2 \text{ kN /m}^2$$

Similarly, the pore water pressure also increases by the same magnitude of 20.2 kN/m^2. Thus, there is no change in effective stress. Any fluctuation in the level of free water above the ground surface would not result in any change in effective stress at any depth within the soil deposit.

Question 10.9: A seepage flow condition is shown in the figure. The saturated unit weight of the soil $\gamma_{sat} = 18 \ kN/m^3$. Using unit weight of water, $\gamma_{wat} = 9.81 kN/m^3$, determine the effective vertical stress (expressed in kN/m^2) on plane X-X.

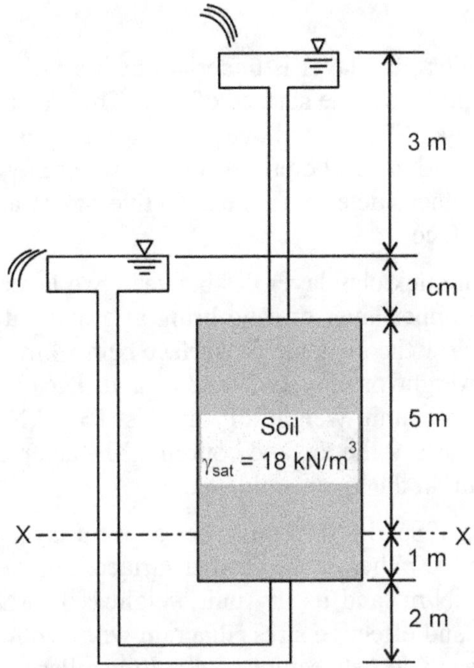

Figure 10.7

Solution: Total loss of head, $h_f = 3$

Total length of seepage, L= 6m

Loss of head for 5 m seepage length

$$= \frac{3}{6} \times 5 = 2.5 \text{ m}$$

Pressure head at X-X is

$$h_p = 9 - 2.5 = 6.5 \text{ m}$$

At plane XX:

Total stress, $\sigma = \gamma_{sat} \times 5 + \gamma_w \times 4$

$$= 18 \times 5 + 9.81 \times 4$$

$$= 129.24 \text{ kPa}$$

Neutral stress, $u = \gamma_w h_p = 9.81 \times 6.5 = 63.765$ kPa

$$\sigma' = \sigma - u = 129.24 - 63.765 = 64.475 \text{ kP.}$$

Questions

Question 1: A 10 thick clay layer is underlain by a sand layer of 20 m depth. The water table is 5 m below the surface of clay. The water table is 5 m below the surface of clay layer. The soil above the water table is capillary saturated. The value of γ_{sat} is 19 kN/m³. The unit weight of water is γ_w. If now the water table rises to the surface, determine the effective stress at a point 10 meter below the ground surface.

Question 2: A 5 mm thick clay layer lies between two layers of sand each 4 m thick, the top of the upper layer of sand bring at ground at ground level. The water table is 2 m below the piezometric surface being 4 m above ground level. The saturated unit weight of clay is 20kN/m³ and that of sand is 19 kN/m³. Above the water table unit weight of sand is 16.5 kN/m³. Calculate the effective vertical stresses at the top and bottom of the layer. Also draw the total vertical stress diagram in the given soil layers.

Question 3: A layer of sand 8·0 m thick lies above a layer of clay. The water table is at a depth of 1·0 m below the ground surface. The saturated unit weight of the sand is 20·0 kN/m³ and its dry unit weight is 17 kN/m³. Plot the total stress, neutral stress and effective stress diagram with depth up to 8·0 m. If the sand above the water table gets saturated due to capillary moisture, what will be the changes in the stress diagram?

Question 4: A layer of sand 6·0 m thick lies above a layer of clay soil. The water table is at a depth of 2·0 m below the ground surface. The void ratio of the sand layer is 0·6 and the degree of saturation of the sand layer above the water table is 40%. The void ratio of the clay layer is 0·7. Determine the total stress, neutral stress and effective stress at a point 10m below the ground surface. Assume specific gravity of the sand and clay soil as respectively as 2·65 and 2·7.

Question 5: A river 5m deep consists of a sand bed with saturated unit weight of 20kN/m³. $\gamma_w = 9.81$ kN/m³. Calculate the effective stress at 5m from the top of sand bed.

Question 6: For the subsoil condition shown in figure 10.8 below (a) what are the effective stress values at 1m, 2m and 4m depths? Assume $\gamma_w = 10$ kN/m³.

Question 7: (a) For the subsoil condition shown in figure 10.9 given below draw the total, neutral and effective stress diagrams upto a depth of 8m. Neglect capillary flow. (b) it the water table rise upto ground surface, what is the change in effective stress at 8m.

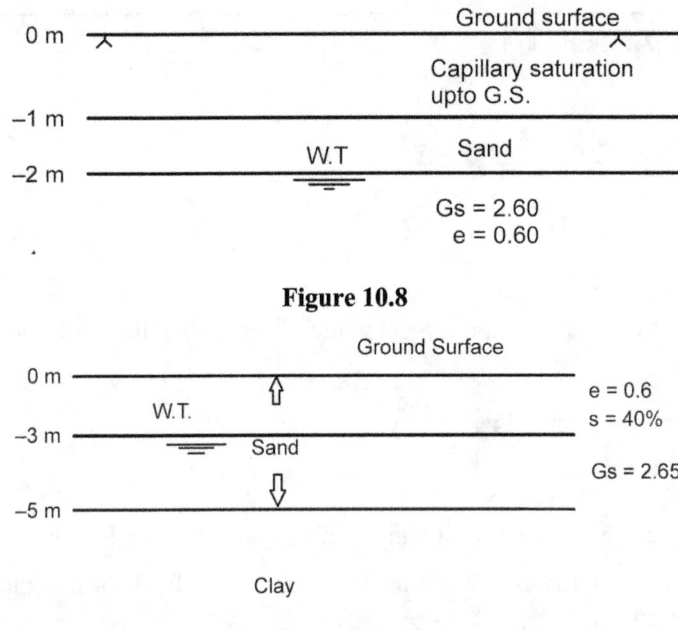

Figure 10.8

Figure 10.9

Question 8: A sand layer of 10·0 m thick lies above a layer of clay. The water table is at a depth of 3·0 m below the ground surface. The saturated unit weight of the sand is 28·0 kN/m³ and its dry unit weight is 14 kN/m³. Draw the total stress, neutral stress and effective stress diagram with depth up to 10·0 m. If the sand above the water table gets saturated due to capillary moisture, what will be the effect in the stress diagram?

Question 9: A sand layer of 12·0 m thick lies above a layer of clay soil. The water table is at a depth of 5·0 m below the ground surface. The void ratio of the sand layer is 0·8 and the degree of saturation of the sand layer above the water table is 35%. The void ratio of the clay layer is 0·8. Determine the total stress, neutral stress and effective stress at a point 10m below the ground surface. Assume specific gravity of the sand and clay soil as respectively as 2·6 and 2·7.

Question 10: A sand layer of 20·0 m thick lies above a layer of clay soil. The water table is at a depth of 8·0 m below the ground surface. The void ratio of the sand layer is 0·9 and the degree of saturation of the sand layer above the water table is 50%. The void ratio of the clay layer is 0·95. Determine the total stress, neutral stress and effective stress at a point 10m below the ground surface. Assume specific gravity of the sand and clay soil as respectively as 2·6 and 2·7.

CHAPTER 11

Stability of Slopes

Question 11.1: Determine the largest value of Taylor's stability number.

Solution:

Taylor's stability number is given by

$$S_n = \frac{c}{\gamma \times H_c}$$

Where, c = cohesion, γ = unit weight of soil, and H_c = safe height

Its maximum value is 0.261 and it occurs for soft clays having angle of internal friction zero

i. e., for $\phi = 0$ soils.

Question 11.2: A 10 m deep canal is constructed in a purely cohesive soil having the following properties: c = 0.2 kg/cm^2, ϕ = 0, G = 2.5, e = 0.5. The stability number is 0.1. In a canal running in full condition, determine the factor of safety w.r.t. cohesion against failure of side slopes.

Solution:

As the canal is running in full condition, the soil will be under the water level or submerged in water.

Hence the submerged density will be used.

Taylor's stability number $(S_n)' = \dfrac{c}{\gamma \times H_c}$

Where,

c = cohesion, ρ_{sub} = submerged unit weight, H = height of canal, F = factor of safety.

Given:

$(S_n)' = 0.1$

$$c = 0.2 \text{ kg/cm}^2 = 200 \text{ g/cm}^2$$

H = 10 m = 1000 cm, G = 2.5, e = 0.5

So, now we need to determine ρ_{sub}

We know,

$$\rho_{sub} = \frac{G-1}{1+e} \times \rho_w = \frac{2.5-1}{1+0.5} \times 1 = 1 \text{ g/cm}^3$$

Now, putting the values in $(S_n)' = \dfrac{c}{\rho_{sub} \times H_c \times F}$ we get,

$$0.1 = 200/(1 \times 1000 \times F)$$

or $\qquad F = 200/(1000 \times 0.1)$

or $\qquad F = 2$

Question 11.3: An embankment in clayey soil of 5 m height is to be constructed using factor of safety of 2.5. It is to be assumed that stability number is 1/45 and unit weight of soil is 18kN/m^3. Find the minimum cohesive strength (in kN/m^2) which the soil should have.

Solution:

Factor of safety $F = \dfrac{H_c}{H}$

Where H_c = critical height and H = depth of cut

$$S_n = \frac{c}{\gamma \times H_c}$$

Where,

S_n = stability number and

γ = unit wt. of soil

$F = 2.5$, $H = 5$ m,

$$S_n = \frac{1}{45}$$

$$\gamma = 18 \text{ kN/m}^3$$

$$H_c = 2.5 \times 5 = 12.5$$

$$S_n = \frac{c}{\gamma \times H_c}$$

$\Rightarrow \qquad c = S_n \times \gamma \times H_c$

$$c = \frac{1}{45} \times 18 \times 12.5 = 5 \text{ kN/m}^2.$$

Question 11.4: A slope is to be constructed at an angle of 30^0 to the horizontal using a soil having the properties, $c = 15$ kN/m^2, $\gamma = 19$ kN/m^3. Taylor's stability number is 0.046. If a factor of safety (with respect to cohesion) of 1.5 is required, then determine the safe height of the slope.

Solution:

Taylor's stability number is given by;

$$S_n = \frac{c}{\gamma \times H_c}$$

Factor of safety with respect to height,

$$F = \frac{H_c}{H}$$

Factor of safety with respect to cohesion,

$$F = \frac{c}{c_m}$$

Where, c_m = mobilised cohesion

c = total cohesion

From the observation, it is found that factor of safety with respect to height is equal to factor of safety with respect to cohesion.

Given,

$$c = 15 \text{ kN/m}^2, \gamma = 19 \text{ kN/m}^3, S_n = 0.046, (FOS)_c = 1.5$$

Taylor's stability number is given by,

$$S_n = \frac{c}{\gamma \times H_c}$$

$$\Rightarrow \qquad S_n = \frac{c}{\gamma \times (FOS)_c \times H}$$

$$0.046 = \frac{15}{19 \times 1.5 \times H}$$

$$H = 11.44 \text{ m}$$

Question 11.5: For 6 m deep excavation in soft clay, $\gamma = 18 \text{ kN/m}^3$, $c = 26$ kN/m^2, Taylor's stability number $S_n = 0.172$. Determine the factor of safety (F_c) against sliding.

Solution:

For purely cohesive soil,

$$S_n = \frac{c}{\gamma \times H_c \times F_c}$$

Where,

c = cohesion, F_c = Factor of safety with respect to cohesion, and H = depth of excavation.

Given:

$$c = 26 \text{ kN/m}^2, \gamma_t = 18 \text{ kN/m}^3, S_n = 0.172, \text{ and } H = 6 \text{ m}$$

Using the formula,

$$S_n = \frac{26}{18 \times 6 \times F_c} = 0.172$$

$$F_c = 1.4$$

Question 11.6: A cutting 8m deep is to be made in a saturated clay soil having $\gamma = 20$ kN/m^3, $c_u = 20$ kN/m^2 and $\phi = 0$. A hard stratum exists at a depth of 12m below the ground level. Determine the angle of the slope at which the failure would occur.

Solution:

$H = 8$m

Therefore, $D_f = 1.5$m

Assume a factor of safety of 1.5.

$$S_n = \frac{c}{\gamma \times H_c \times F_c}$$

$$S_n = \frac{20}{20 \times 8 \times 1.5} = 0.083$$

From Taylor's stability chart

For $S_n = 0.083$ and $D_f = 1.5$

$$i = 8^0$$

Hence for a factor of safety of 1.5, the failure would occur at an angle of 8^0.

Question 11.7: A long natural slope of cohesionless soil is inclined at 15^0 to the horizontal. Taking $\phi = 300$, determine the factor of safety of the slope. If the slope is completely submerged, what will be the change in the factor of safety?

Solution:

Given

$$\phi = 30^0$$

$$i = 15^0$$

$$F = \frac{\tau_c}{\tau} = \frac{\tan\phi}{\tan i} = \frac{\tan 30^0}{\tan 15^0} = 2.15$$

Effect of submergence

$$F = \frac{\tau_c}{\tau} = (\gamma' \times z \times \cos^2 i) \times \frac{\tan\phi}{\gamma' \times z \times \sin i \times \cos i} = \frac{\tan\phi}{\tan i} = \frac{\tan 30^0}{\tan 15^0} = 2.15$$

Thus, the factor of safety will remain the same, except that ϕ is to be determined under submerged condition.

Question 11.8: In order to find a factor of safety of a slope of an earth dam during steady seepage, the section of the dam was drawn to a scale of 1:4 and the following results were obtained on a trial slip circle.

Area of N-rectangle = 10cm^2

Area of T-rectangle = 8cm^2

Area of U-rectangle = 4.8cm^2

Length of the failure arc = 12cm

The laboratory test on slope material produced c = 18 kN/m^2 and $\phi = 25^0$. Determine the factor of safety of slope of unit weight of soil is 20 kN/m^3.

Solution:

Given scale 1cm = 4cm

$$\gamma = 20 \text{ kN/m}^3$$

$$c = 18 \text{ kN/m}^3$$

$$\phi = 25^0$$

Given scale 1cm = 4cm, therefore x = 4

Consider 1cm length of the dam

$$\Sigma N = A_N x^2 \gamma = 10 \times (4)^2 \times 20 = 3200 \text{kN}$$

$$\Sigma T = A_T x^2 \gamma = 8 \times (4)^2 \times 20 = 2560 \text{kN}$$

$$\Sigma U = A_U x^2 \gamma_w = 4.8 \times (4)^2 \times 9.81 = 753.4 \text{kN}$$

$$L' = 12 \times 4 = 48 \text{m}; \ c' = 18 \text{ kN/m}^2$$

$$F = \frac{c' \times L' + \tan\varphi \times \Sigma(N - U)}{\Sigma T}$$

$$= \frac{18 \times 48 + \tan 25^0 \times (3200 - 753.4)}{2560}$$

$$= 0.78$$

Question 11.9: A long slope is formed in a soil with shear strength parameter: c' = 0 and $\phi' = 34°$. Firm stratum lies below the slope and it is assumed that water table may occasionally rise to the surface, with seepage taking place parallel to the slope. Use $\gamma_{sat} = 18$ kN/m^3 and $\gamma_w = 10$ kN/m^3. Determine the maximum slope angle (in degrees) to ensure the factor of safety 1.5, assuming a potential failure surface parallel to the slope.

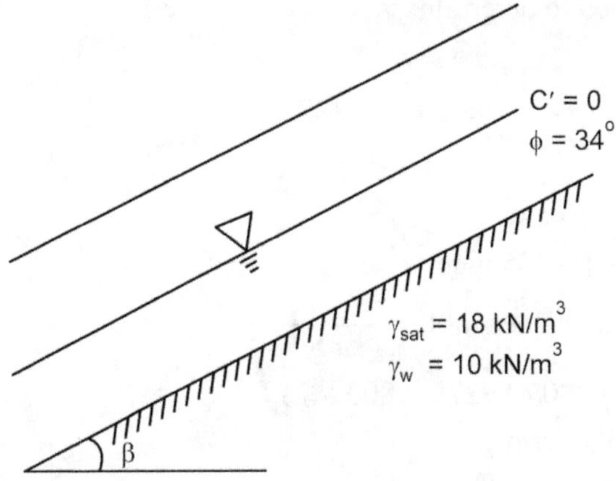

Figure 11.1

Solution:

Factor of safety when water surface rises to surface is given by

$$F.O.S = \frac{\gamma_{sub}}{\gamma_{sat}} \times \frac{\tan\phi}{\tan\beta}$$

$$\tan\beta = \frac{\gamma_{sub}}{\gamma_{sat}} \times \frac{\tan\phi}{F.O.S.}$$

$$\tan\beta = \frac{(18-10)}{(18 \times 1.5)} \times \tan 34°$$

$$\Rightarrow \beta = 11.30^0$$

Question 11.10: An 8 m deep canal is constructed in purely cohesive soil having following properties: c = 0·2 kg/cm^2, ϕ = 0°, G = 2·5 and e = 0·6. The stability number is 0·2. For the canal running in full condition, determine the factor of safety w.r.t. cohesion against failure of side slopes.

Solution:

Given:

$$S = 0.05$$

$$c = 0.2 \text{ kg/cm}^2$$

$$= 200 \text{ g/cm}^2$$

$$H = 8 \text{ m}$$

$$= 800 \text{ cm,}$$

$$G = 2.5,$$

$$e = 0.6$$

So, now we need to determine γ_{sub}

We know,

$$\rho_{sub} = \frac{G-1}{1+e} \times \rho_w$$

$$= \frac{2.5-1}{1+0.6} \times 1$$

$$= 0.9375 \text{ g/cm}^3$$

Now, putting the values in $S_n = \dfrac{c}{\rho_{sub} \times H_c \times F}$

or $0.2 = 200/(0.9375 \times 800 \times F)$

or $F = 0.2667/0.2$

or $F = 1.33$

Question 11.11: What is the critical height of the slope of infinite extent having a slope angle 25°, if it is made of clay having $c = 30 \text{ kN/m}^2$, $\phi = 20°$, $e = 0.65$ and $G = 2.7$, when the slope is submerged?

Solution:

For infinite slope when the slope is submerged,

$$\text{FOS} = \frac{c + \gamma_{sub} z \cos^2 \beta \tan \phi}{\gamma_{sub} \, z \, \cos \beta \sin \beta}$$

Where,

c = cohesion (kN/m^2)

β = slope angle (degrees)

ϕ = angle of internal friction (degrees)

γ_{sub} = the submerged unit weight of soil (kN/m^3)

At FOS = 1,

$z = H_c$

$$H_c = \frac{c}{\gamma_{sub} \cos \beta (\sin \beta - \cos \beta \tan \phi)}$$

Given, $\beta = 25°$

$\qquad c = 30 \text{ kN/m}^2$,

$\qquad \phi = 20°$,

$e = 0.65$ and

$G = 2.7$,

$\qquad \gamma_w = 9.81 \text{ kN/m}^3$

$$\gamma_{sub} = \left(\frac{G-1}{1+e}\right) \times \gamma_w$$

$$\gamma_{sub} = \left(\frac{2.7-1}{1+0.65}\right) \times 9.81$$

$$= 10.107 \text{ kN/m}^3$$

Critical height,

$$H_c = \frac{3}{10.107 \times \cos 25° (\sin 25° - \cos 25° \tan 20°)}$$

$$H_c = 35.31 \text{ m}$$

Question 11.12: An unsupported slope of height 15 m is shown in the figure (not to scale), in which the slope face makes an angle 50° with the horizontal. The slope material comprises purely cohesive soil having undrained cohesion 75 kPa. A trial slip circle KLM, with a radius 25 m, passes through the crest and toe of the slope and it subtends an angle 60° at its center O. The weight of the active soil mass (W, bounded by KLMN) is 2500 kN/m, which is acting at a horizontal distance of 10 m from the toe of the slope. Consider the water table to be present at a very large depth from the ground surface.

Considering the trial slip circle KLM, determine the factor of safety against the failure of slope under undrained condition.

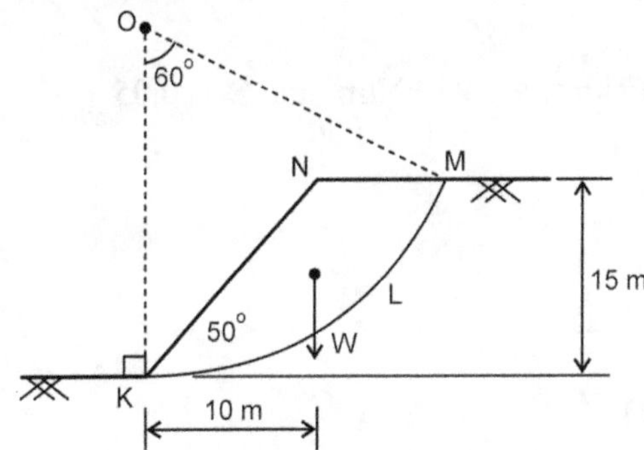

Figure 11.2

Solution:

$$FOS = \frac{\text{Resisting moment}}{\text{Actuating moment}}$$

$$FOS = \frac{c_u \times l \times R}{w \times \bar{x}}$$

l = Length of arc KLM

$\bar{x}$ = distance of weight 'w' from toe

$$FOS = 75 \times 2\pi \times 25 \times \frac{60}{360} \times \frac{25}{2500 \times 10}$$

or $$FOS = 1.96$$

Question 11.13: For a clay slope of height 20 m, the stability number is 0.05, unit weight of soil $\gamma = 25$ kN/m^3, and $c = 30$ kN/m^2, find the critical height of slope.

Solution:

Taylor stability number method:

(i) Taylor introduced a dimensional parameter, called Taylor's stability number, which is given by

$$S_n = \frac{c}{\gamma Hc} = \frac{c}{\gamma F H}$$

Where,

H= effective height of slope

H_c = critical height of slope

$$F = FOS$$

(ii) In this methid, Taylor's stability number is read from Taylor's chart in order to analyze the stability of slope for given value of strength parameters of the soil.

(iii) Taylor's method is suitable for c-ϕ soils and for $\phi = 0$(pure clays)

Given,

$$c = 30 \text{ kN/m}^2, \gamma = 25 \text{ kN/m}^3 \text{ and } S_n = 0.05 \frac{c}{\gamma Hc}$$

$\because$ We know that

$$S_n = \frac{c}{\gamma Hc}$$

$$\Rightarrow 0.05 = \frac{30}{25 \times Hc}$$

or $\qquad$ $H_c = 24$ m

Question 11.14: The soil profile above the rock surface for a 25° infinite slope is shown in the figure, where s_u is the untrained shear strength and γ_t is total unit weight.

Determine the depth at which slip will occur.

Solution: If the failure occurs in top layer,

$$\tau = \gamma_t h \times cosi \times sini$$

$$= h \times \cos 25° \times \sin 25° \times 16$$

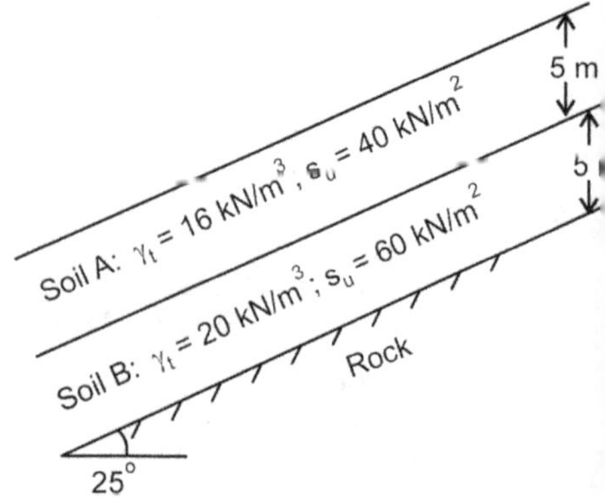

Figure 11.3

Untrained shear strength $S_u = 40$ kN/m^2

$\tau > S_u$

Therefore,

$$h \times cos\,25° \times sin\,25° \times 16 > 40$$

or $h > 6.52$

But for top soil layer, the maximum depth h is 5 metres.

Therefore, failure occurs in the bottom layer.

$$\tau = [\gamma_{t1}h_1 \times cos\,i + \gamma_{t2}h_2 \times cos\,i] \times sin\,i$$

$$= 5\,cos\,25° \times 16 + h_2cos\,25° \times 20]\,sin\,25°$$

Untrained shear strength $S_u = 60$ kN/m^2

$$\tau = S_u$$

Therefore,

$$[5\,cos\,25° \times 16 + h_2cos\,25° \times 20]\,sin\,25° = 60$$

or $h_2 = 3.83\,m$

From top, the depth of slip $= 5 + 3.83 = 8.38$ m.

Question 11.15: If there is a sudden drawdown of water in the canal and if Taylor's stability number for the reduced value of ϕ_w is 0.126, determine the factor of safety with respect to cohesion against the failure of bank slopes. Given $c_u = 12$ kN/m^2, $\gamma' = 8.09$ kN/m^3 and height of canal bank $= 10$ m.

Solution: Even if there is a sudden drawdown of water in the canal, the soil remains saturated for some time. Hence, unit weight of the soil will be the saturated unit weight i.e.

$$\gamma_{sat} = \gamma' + \gamma_w$$

$$= 8.09 + 9.81$$

$$= 17.9\text{ kN/m}^3$$

Since,

Taylor's stability number,

$$S_n = c_u / F_c \times \gamma_{sat} \times H$$

$$F_c = c_u / S_n \times \gamma_{sat} \times H$$

$$= 12/0.126 \times 17.9 \times 10$$

$$= 0.53$$

Question 11.16: An infinitely long slope is made up of a c- ϕ soil having the properties: cohesion, c = 20 kPa, and dry unit weight, γ_d = 16 kN/m³. The angle of inclination and critical height of the slope are 40° and 5 m, respectively. To maintain the limiting equilibrium, determine the angle of internal friction of the soil.

Solution: Given,

$$c = 20 \text{ kPa,}$$

$$\gamma_d = 16 \text{ kN/m}^3$$

$$i = 40°,$$

$$H = 5m$$

$$H_c = c/(\gamma_d \, (\tan i - \tan \phi)\cos^2 i)$$

$$5 = 40 \, / \, (16 \, (\tan 40° - \tan \phi) \, \cos^2 40°$$

$$\tan \phi = 0.413$$

$$\phi = 22.44°.$$

Question 11.17: An infinite slope is to be constructed in a soil. The effective stress strength parameters of the soil are c′ = 0 and $\phi′$ = 30°. The saturated unit weight of the slope is 20 kN/m³ and the unit weight of water is 10 kN/m³. Assuming that seepage is occurring parallel to the slope, determine the maximum slope angle for a factor of safety of 1.5.

Solution:

$$F = (\gamma' \times \tan \phi') \, / \, (\gamma_{sat} \times \tan i)$$

$$\gamma' = \gamma_{sat} - \gamma_w$$

$$(20 - 10) = 10 \text{ kN/m}^3$$

$$1.5 = (10 \times \tan 30°) \, / \, (20 \times \tan i)$$

$$\tan i = (\tan 30°) \, / \, (1.5 \times 2) = 0.192$$

$$i = 10.89°.$$

Questions

Question 1: A long slope is formed in a soil with shear strength parameters: c′ = 0 and ϕ ′= 34°. A firm stratum lies below the slope and it is assumed that the water table may occasionally rise to the surface, with seepage taking place parallel to the slope. Use γ_{sat} = 18 kN/m³ and γ_w = 10 kN/m³. Determine the maximum slope angle (in degrees) to ensure a factor of safety of 1.5, assuming a potential failure surface parallel to the slope.

Question 2: The soil profile above the rock surface for a 25° infinite slope is shown in the figure 11.4, where s_u is the undrained shear strength and γ_t is total unit weight. At what depth the slip will occur?

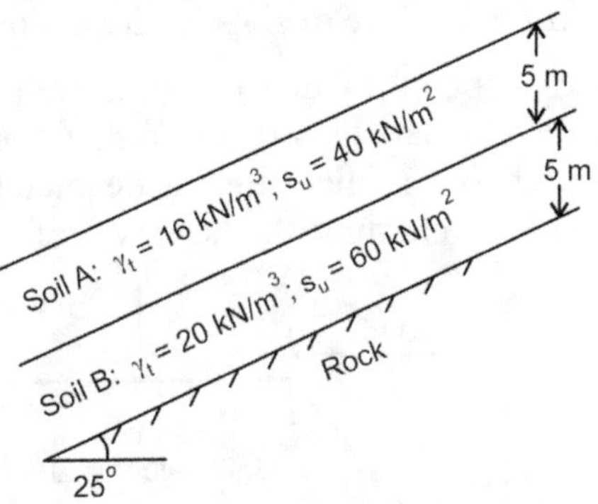

Figure 11.4

Question 3: Determine the factor safety of an infinite soil slope shown in the figure 11.5 having the properties c = 0, ϕ = 35°, γ_{dry} = 16 kN/m^3, and γ_{sat} = 20 kN/m^3.

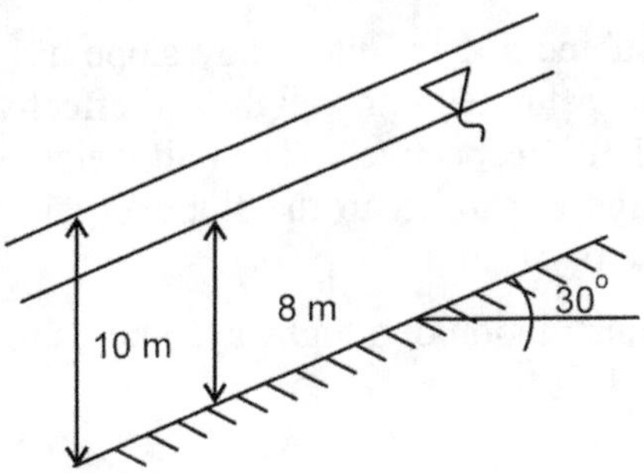

Figure 11.5

Question 4: An unsupported slope of height 15m is shown in the figure 11.6 (not to scale) in which the slope face makes an angle 50° with the horizontal. The slope material comprises purely cohesive soil having undrained cohesion 50kPa. A trail slip circle KLM with a radius 25m, passes through crest and toe of the slope and it subtend an angle 60° at its center O. The weight of the active soil mass (W, bounded by

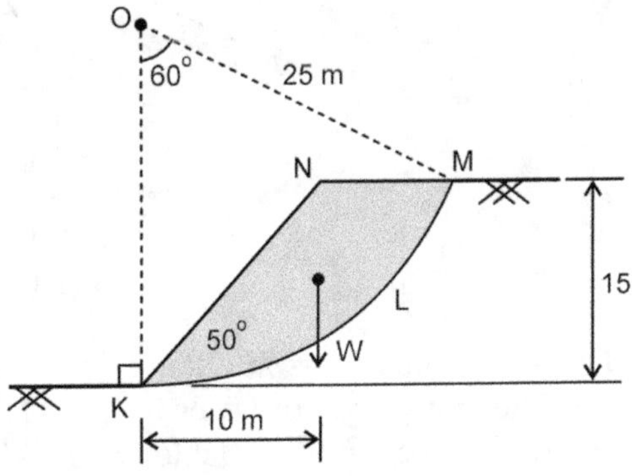

Figure 11.6

KLMN) is 2700 kN/m, which is acting at a horizontal distance of 10m from the toe of the slope. Consider the water table to be present at a very large depth from ground surface. Considering the trail slip circle KLM, determine the factor of the safety against the failure of slope under undrained condition.

Question 5: A 10m high slope of dry clay soil (unit weight = 20kN/m³) with a slope angle of 45° and the circular slip surface, is shown in the figure 11.7 (not drawn to scale). The weight of the slip wedge is denoted by W. The undrained unit cohesion (c_u) is 60 kPa. Determine the factor of safety of the slope against slip failure.

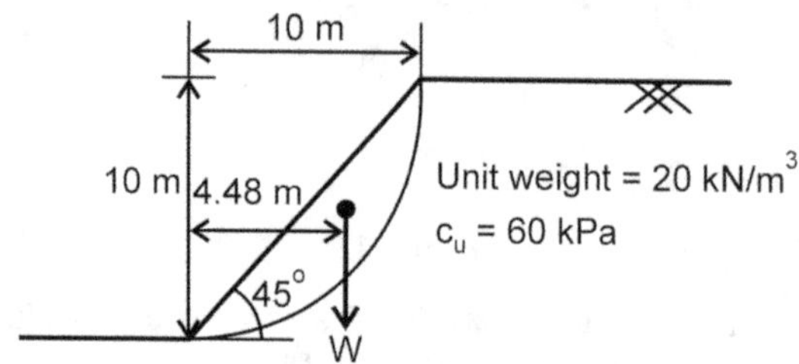

Figure 11.7

Question 6: A fully submerged infinite sandy slope has an inclination of 30° with the horizontal. The saturated unit weight and effective angle of friction of sand are 18 kN/m³ and 38°, respectively. The unit weight of water is 10 kN/m³. Assume that the seepage is parallel to the slope. Against shear failure of the slope, determine the factor of safety.

Question 7: For the construction of a highway, a cut is to be made as shown in the figure 11.8.

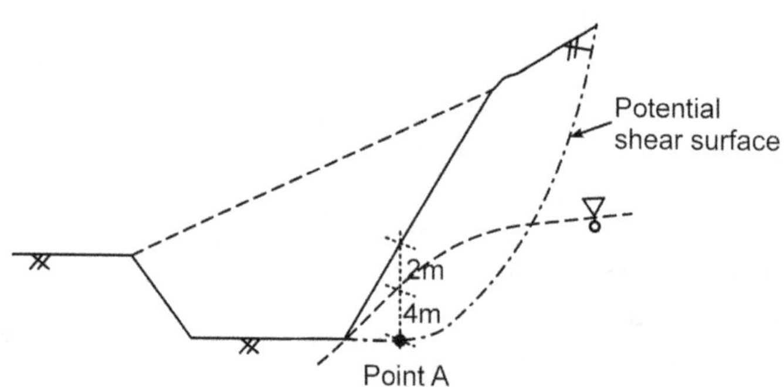

Figure 11.8

The soil exhibits $c' = 20$ kPa, $\phi' = 18°$ and the undrained shear strength = 80kPa. The unit weight of water is 9.81 kN/m³. The unit weight weights of the soil above and below water table are 18 kN/m³ and 20 kN/m³, respectively, If the shear stress at point A is 50 kPa, determine the factors of safety against the shear failure at this point, considering the undrained and drained conditions, respectively.

Question 8: A canal having a side slope 1:1 is proposed to be constructed in a cohesive soil to a depth of 10m below the ground surface. The soil properties are ϕ_u=15°, c_u = 12kPa and G = 2.65.If Taylor's stability number S_n is 0.08 and canal flows full, determine the factor of safety with respect to cohesion against failure of the canal bank slopes.

Question 9: An infinite soil slope with an inclination of 35° is subjected to seepage parallel to its surface. The soil has $c' = 100$ kN/m² and $\phi' = 30°$. Using the concept of mobilized cohesion and friction, determine the mobilized frictional angleat a factor of safety of 1.5 with respect to shear strength.

Question 10: For the trial slip circle shown in the figure 11.9 with W = 380 kN, the soil properties are: unit weight of soil $\gamma = 18.8$ kN/m³, $\phi = 0°$ and c = 27 kN/m³. Determine the value of restoring moment (kN-m) for the slope shown in the figure.

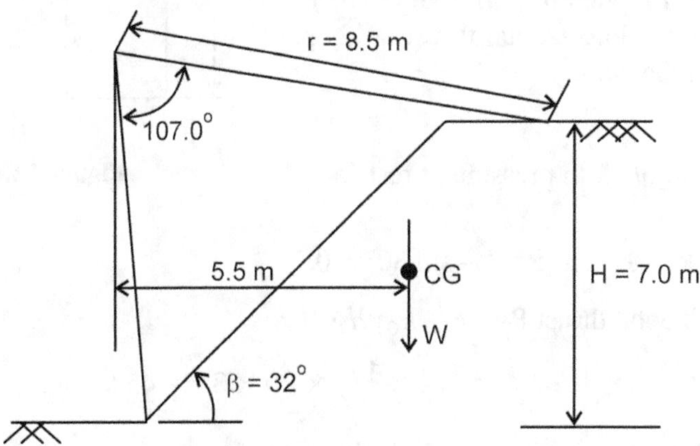

Figure 11.9

Question 11: A slope of 1V:2H is to be made in silty clay having an angle of internal friction of 10° and cohesion of 30 kN/m². The unit weight of soil is 18 kN/m³ and depth of cut is 7.0m. The value of net normal and tangential force is 2741 kN and 914 kN, respectively. Angle between starting point, origin and exit point of the slip surface is 109°. Lever arm distance between origin and exit point of the slip surface is provided as 12 m. Determine factor of safety of the slope using Swedish circle method.

Question 12: Stability analysis by the method of slices for 1:1 slope on the critical slip gave the following results:

Sum of tangential forces = 150 kN; sum of normal forces = 320 kN

Sum of neutral forces = 50 kN; length of failure surface = 18 m

Effective angle of shearing resistance = 15°, effective cohesion = 20 kN/m²

Calculate the factor safety with respect to the shear strength.

CHAPTER 12

Earth Pressure

Question 12.1: A 3 m high vertical earth retaining wall retains a dry granular backfill with the angle of internal friction of 30° and unit weight of 20 kN/m³. If the wall is prevented from yielding (no movement), determine the total horizontal thrust in kN per unit length of the wall.

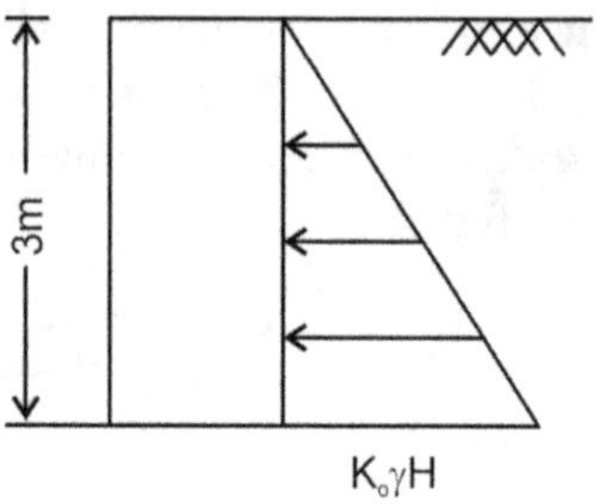

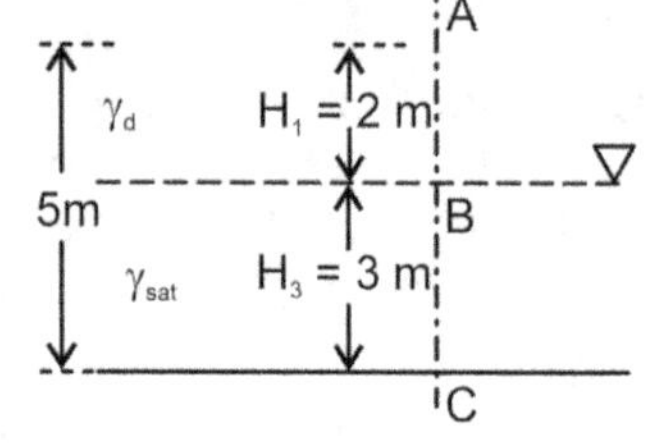

Figure 12.1

Solution:

The coefficient of earth pressure at rest for dry sand is:

$$k_0 = 1 - \sin\phi = 1 - \sin 30° = 0.5$$

Total horizontal thrust $P_0 = {}^1/_2\, k_0\gamma H \cdot H$

$$= {}^1/_2 \times {}^1/_2 \times 20 \times 3^2 = 45 \text{ kN/m.}$$

Question 12.2: A cohesionless soil with a void ratio of e = 0.6 and specific gravity of soil solids G_S = 2.65 exists at a site where the water table is located at a depth of 2 m below the ground surface. Assuming a value of the coefficient of earth pressure at rest K_o = 0.5, calculate the following quantities at a depth of 5 meters below the ground surface (figure 12.2):

(a) total stress q and effective stress q'

(b) Vertical pressure and horizontal pressure

(c) pore water pressure u.

Figure 12.2

Solution:

Given: e = 0.6; G = 2.65; K_0 = 0.5

$$\gamma_d = G\gamma_w/(1+e) = 2.65 \times 9.81/(1 + 0.6)$$

$$= 16.248 \text{ kN/m}^3;$$

$$\gamma_{sat} = (G+e)\gamma_w/(1+e) = (2.65 + 0.6)\times 9.81/(1 + 0.6) = 19.927 \text{ kN/m}^3;$$

$$\gamma' = 19.927 - 9.81 = 10.117 \text{ kN/m}^3;$$

(a) Vertical pressure:

Now,

$$\sigma = \sigma_z = \gamma_d H_1 + \gamma' H_2$$

$$= 16.248 \times 2 + 10.117 \times 3 = 62.487 \text{ kN/m}^3$$

$$\sigma_v = \sigma_z = \gamma_d H_1 + \gamma_{sat} H_2 = 16.248 \times 2 + 19.927 \times 3 = 92.277 \text{ kN/m}^2$$

(b) Pore pressure:

$$u = \sigma_u = \gamma_w H_2 = 9.81 \times 3 = 29.43 \text{ kN/m}^3$$

(c) Horizontal pressure:

$$= \text{earth pressure at rest} = p_0 = K_0 \gamma_d H1 + K_0 \gamma' H_2$$

$$= 0.5 \times 16.248 \times 2 + 0.5 \times 10.117 \times 3 = 31.424 \text{ kN/m}^2$$

$$u_h = \gamma_w H_2 = 9.81 \times 3 = 29.43 \text{ kN/m}^2$$

Total pressure $= p_0 + u_h = 31.424 + 29.43 = 60.854 \text{ kN/m}^2$.

Question 12.3: Compute the intensity of passive earth pressure at a depth of 8 m in cohesionless sand with an angle of internal friction of 30° when the water rises to the ground level. The saturated unit weight of sand is 21 kN/m^3 and γ_w = 9.81 kN/m^3.

Solution:

Here, $\quad \gamma_{sat} = 21 \text{ kN/m}^3 \; ; \; \phi = 30°$

$$K_p = (1 + \sin \phi) / (1 - \sin \phi)$$

$$= (1 + \sin 30°) / (1 - \sin 30°) = 3;$$

$$\gamma' = 21 - 9.81 = 11.19 \text{ kN/m}^2$$

$$P_p = K_p \gamma' H + \gamma_w H = 3 \times 11.19 \times 8 + 9.81 \times 8 = 347.04 \text{ kN/m}^2.$$

Question 12.4: A counterfort wall of 10 m height retains non-cohesive backfill. The void ratio and angle of internal friction of the backfill respectively are 0.7 and 30° in the loose state and they are 0.40 and 40° in the dense state. Calculate and compare active and passive earth pressure in both the states. Take specific gravity of soil grains as 2.7?

Solution:

Loose state:

$$e = 0.70, \phi = 30°$$

$$G = 2.7$$

$$K_a = (1-\sin 30°)/(1+\sin 30°) = 1/3$$

$$K_p = 3$$

$$\gamma_d = G \times \gamma_w/ (1+e)$$
$$= 2.7 \times 9.81/(1+0.7)$$
$$= 15.58 \text{ kN/m}^2$$

$$P_a = 0.5 \, K_a\gamma \, H^2 = 0.5 \times 1/3 \times 15.58 \, (10)^2$$
$$= 259.67 \text{ kN/m}$$

$$P_p = 0.5 K_p\gamma \, H^2 = 0.5 \times 3 \times 15.58 \, (10)^2$$
$$= 2337 \text{ kg/m}$$

Dense state:

$$e = 0.40 \text{ and } \phi = 40°; \, G = 2.7$$

$$K_a = (1-\sin 30°)/(1 + \sin 30°) = 0.2174$$

$$K_p = 1/ \, 0.2174 = 4.5989.$$

$$\gamma_d = G\gamma_w/(1+e) = 2.7 \times 9.81 \, /(1+0.4)$$
$$= 18.919 \text{ kN/m}^3$$

$$P_a = 0.5 \, K_a\gamma \, H^2 = 0.5 \times 0.2174 \times 18.919 \, (10)^2 = 205.65 \text{ kN/m}$$

$$P_a = 0.5 \, K_p\gamma \, H^2 = 0.5 \times 4.5989 \times 18.919 \, (10)^2 = 4350.4 \text{ kN/m}$$

Question 12.5: A circular raft foundation of 20 m diameter and 1.6 m thick is provided for a tank that applies a bearing pressure of 110 kPa on sandy soil with Young's modulus, $E_s' = 30$ MPa and Poisson's ratio, $v_s = 0.3$. The raft is made of concrete ($E_c = 30$ GPa and $v_c = 0.15$). Considering the raft as rigid, determine the elastic settlement of the raft.

Solution:

Immediate settlement of cohesive soils:

The linear theory of elasticity is used to determine the elastic settlement S_i of the footings on saturated clay

$$S_i = qB \left(\frac{1-\mu^2}{Es}\right) I$$

Where,

$\quad$ q = uniformly distributed load (kPa),

$\quad$ B = characteristic length of the loaded area

$\quad$ E_S = modulus of elasticity of the soil (MPa),

$\quad$ μ = Poisson's ratio

I = Influence factor

= Influence factor for rigid footing = 0.8

Given:

q = 110 kPa, E_S = 30 MPa

μ = 0.3, B = 20 m

I for rigid footing = 0.8

$$S_i = qB\left(\frac{1-u^2}{Es}\right)I$$

$$S_i = 110 \times 20\left(\frac{1-0.3^2}{30}\right)0.8$$

$$S_i = 53.38 \text{ mm}$$

Question 12.6: A rigid smooth retaining wall of height 7m with vertical backface retains saturated clay as backfill. The saturated unit weight and undrained cohesion of the backfill are 17.2 kN/m³ and 20 kPa respectively. Determine the difference in the active lateral forces on the wall (in kN per meter length of wall), before and after occurrence of tension cracks.

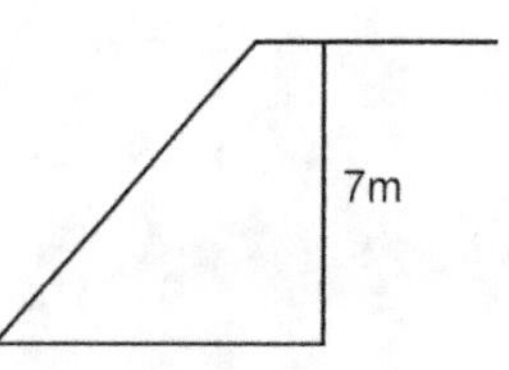

Figure 12.3

Solution:

ϕ = 0, γ = 17.2 kN/m³,

c = 20 kPa

K_a = 1

$2c(k_a)^{1/2} = 2\times 20 = 40$ kPa

$K_a\gamma H - 2c(k_a)^{1/2} = 1 \times 17.2 \times 7 - 2 \times 20$

= 80.4 kPa

Before the occurrence of tension crack

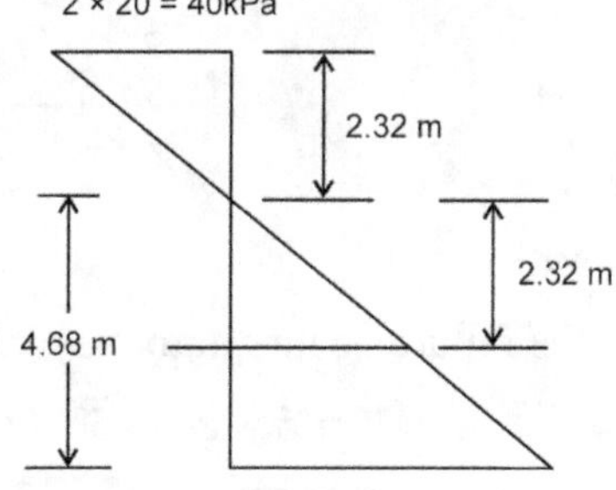

Figure 12.4

$$z_c = 2c/\gamma(k_a)^{1/2}$$

$$= 2 \times 20/17.2\times 1$$

$$= 2.32\text{m}$$

Before formation of tension crack

$$F_a = \tfrac{1}{2}\,k_a\gamma H^2 - 2c(k_a)^{1/2}H$$

$$= \tfrac{1}{2} \times 1 \times 17.2 \times 7^2 - 2 \times 20 \times (1) \times 7$$

$$= 141.4\text{kN/m}$$

After formation of tension crack

$$F_a = \tfrac{1}{2}\,k_a\gamma H^2 - 2c(k_a)^{1/2}H + 2c^2/\gamma$$
$$= \tfrac{1}{2}\times 1\times 17.2\times 7^2 - 2\times 20\times(1)\times 7 + 2\times 20^2/17.2$$
$$= 421.4 - 280 + 46.52$$
$$= 187.92\,\text{kN/m}$$

Difference $= 187.92 - 141.2$

$$= 46.52 \text{ kN/m}$$

Hence, difference in active lateral forces on the wall before and after occurrence of tension cracks = 46.52kN/m.

Question 12.7: The soil profile above the rock surface for a 25° infinite slope is shown in the Figure 12.5, where s_u is the untrained shear strength and γ_t is the total unit weight. Determine the depth at which slip will occur along the slope.

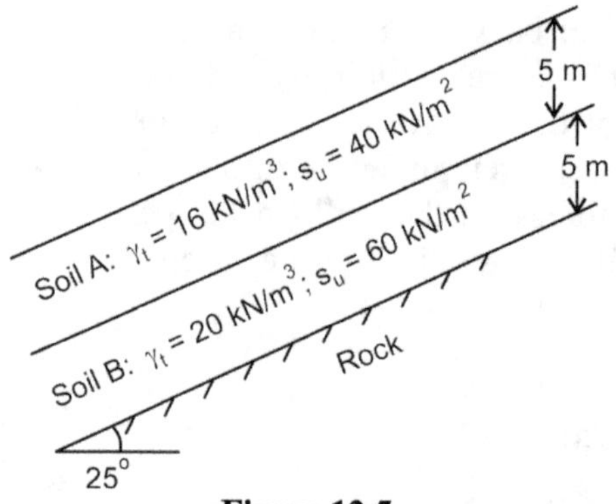

Figure 12.5

Solution:

If the failure occurs in to player, then

$$T = \sigma \times cos\,25° \times sin\,25° \times 16$$
$$S = 40$$
$$T > S$$

Therefore,

$$\Sigma \times cos\,25° \times sin\,25° \times 16 > 40$$
$$\sigma > 6.52$$

But maximum h is 5metres.

Therefore, failure occurs in the bottom layer.

$$T = [5cos25°\times16 + \sigma cos25°\times20]sin25°$$

$$S = 60$$

$$T = S$$

Therefore,

$$[5 \ cos \ 25° \times 16 + \sigma \ cos \ 25° \times 20] \ sin \ 25° = 60$$

$$H = 3.83m$$

From top, depth $= 5 + 3.83$

$$= 8.83 \ m.$$

Question 12.8: Determine the surcharge loading required to be placed on the horizontal backfill of a smooth retaining vertical wall so as to completely eliminate the tensile crack.

Solution:

Active earth pressure for c -ϕ soils

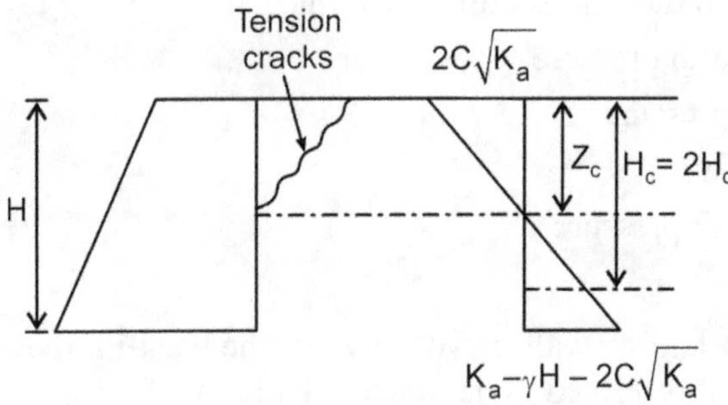

Figure 12.6

Where,

Z_C = Depth of tension zone or depth of tension crack

H_C = Critical height or maximum depth of the unsupported vertical trench

K_a = Coefficient of active earth pressure

P_a = Active earth pressure

$$P_a = K_a\sigma_v - 2c\sqrt{Ka}$$

In the figure, at the top of a trench

$$\sigma_v = 0,$$

$$Pa = -2c\sqrt{K_a}$$

$$K_aq = 2c\sqrt{K_a}$$

$$q = 2c/\sqrt{K_a}$$

Question 12.9: If σ_h, σ_v, $\sigma_{h'}$ and $\sigma_{v'}$ represent the total horizontal stress, total vertical stress, effective horizontal stress and effective vertical stress on a soil element, respectively, determine the coefficient of earth pressure at rest.

Solution:

σ_h: total horizontal stress

σ_v: total vertical stress

$\sigma_{h'}$: effective horizontal stress

$\sigma_{v'}$: effective vertical stress

The coefficient of earth pressure at rest is the ratio of intensity of earth pressure at rest to the vertical stress at a specified depth.

$$K = \frac{\sigma_h}{\sigma_v}$$

Question 12.10: When a retaining wall moves away from the backfill, the pressure exerted on the wall is correctly termed as:

  (a) passive earth pressure

  (b) swelling pressure

  (c) pore pressure

  (d) active earth pressure

Solution:

  (a) When a retaining wall moves towards the backfill, the pressure exerted on the wall is termed as passive earth pressure.

  (b) Swelling pressure is the pressure exerted by the soil on the overlying structure if free swelling of the soil is restrained by the placement of a structure over the soil.

  (c) Pore water pressure is the pressure exerted due to the pore water. It is equal to the product of the depth at which pore water pressure is required and unit weight of water.

  (d) When a retaining wall moves away from the backfill, the pressure exerted on the wall is termed as active earth pressure.

Hence, the correct option is (d).

Question 12.11: A building is to be supported on a reinforced concrete raft covering an area of 14 x 21 metres. The subsoil is clay with an unconfined compressive strength of 84 kN/m². The pressure on the soil due to weight of the building and loads it will carry will be 120 kN/m² at the base of the raft. If the unit weight of excavated soil is 15 kN/m³; at what depth should the bottom of the raft be placed to provide a factor of safety of 3?

Solution:

$$q_u = 84 \text{ kN/m}^2$$

$$c_u = q_u/2 = 42 \text{ kN/m}^2$$

Using Skemptons' equation for ultimate bearing capacity of raft

$$q_{nf} = 5\, c_u\, (1 + 0.2 \times BL)\, (1 + 0.2\, DB) \qquad ....(1)$$

$$q_{nf} = 5 \times 42\, (1 + 0.2 \times 14 \times 21)\, (1 + 0.2\, D \times 14)$$

$$= 2383\, (1 + 0.01429\, D)$$

$$q_s = q_{nf}F + \gamma\, D$$

$$= 2383\, (1 + 0.01429\, D) + 15\, D = 79.33 + 16.134\, D \qquad ....(2)$$

Actual load in density, $q_a = 120 \text{ kN/m}^2$

Equating (2) and (3), we get: $79.33 + 16.134\, D = 120$

From which: $D = 2.52 \text{ m}$

Check:

$$5\, (1 + 0.2\, DB) = 5\, (1 + 0.2\ 2.5214) = 5.18 < 7.5$$

Hence, OK.

Question 12.12: A counterfort wall of 10 m height retains non-cohesive backfill. The void ratio and the angle of internal friction of the backfill respectively are 0.7 and 30° in the loose state and they are 0.40 and 40° in the dense state. Calculate and compare active and passive earth pressure in both the states. Take the specific gravity of soil grains as 2.7.

Solution:

(a) Loose state:

$e = 0.70$

$\phi = 30°$

$G = 2.7$

$$K_a = (1 - \sin 30°)/(1 + \sin 30°) = 1/3$$

$K_p = 3$

$$\gamma_d = G\gamma_w\,/(1+e) = 2.7 \times 9.81/(1+0.7) = 15.58 \text{ kN/m}^3$$

$$P_a = 1/2\, K_a\, \gamma\, H^2 = 1/2 \times 13 \times 15.58\, (10)^2 = 259.67 \text{ kN/m}$$

$$P_p = 1/2\, K_p\, \gamma\, H^2 = 1/2 \times 3 \times 15.58\, (10)^2 = 2337 \text{ kN/m}$$

(b) Dense state:

$e = 0.40$

$\phi = 40°$

$G = 2.7$

$$K_a = (1-\sin 40°)/(1+\sin 40°) = 0.2174$$

$$K_p = 1/0.2174 = 4.5989$$

$$\gamma_d = G\gamma_w (1+e) = 2.7 \times 9.81/(1+0.4) = 18.919 \text{ kN/m}^3$$

$$P_a = 1/2 \; K_a \; \gamma \; H^2 = 1/2 \times 0.2174 \times 18.919 \; (10)^2 = 205.65 \text{ kN/m}$$

$$P_p = 1/2 \; K_p \; \gamma \; H^2 = 1/2 \times 4.5989 \times 18.919 \; (10)^2 = 4350.4 \text{ kN/m}$$

Thus, compaction of backfill results in decrease in active pressure and increase in passive pressure.

Question 12.13: An unsupported excavation is made to the maximum possible depth in a clay soil having $\gamma=18\text{kN/m}^3$, c $=100\text{kN/m}^2$, $\phi=30°$. Determine the active earth pressure according to Rankine's theory at the base level of excavation.

Solution:

The excavation is unsupported which means the maximum possible depth up to which the clay soil can be excavated is $2Z_C$.

The active earth pressure at the depth is,

$$P_a = 2c\sqrt{K_a}$$

where $K_a= (1 - \sin\phi)/(1+\sin\phi)$

$$P_a = 2 \times 100\sqrt{[(1 - \sin30°)/(1+\sin30°)]}$$

$$= 115.47\text{kN/m}^2$$

Alternatively;

$$Z_c = 2c/\sqrt{K_a}$$

Since,

Active earth pressure at depth 2 Z_C

$$Pa = K_a\sigma_v - 2c \sqrt{K_a}$$

$$= K_a \times \gamma \times 2 \; Z_c - 2c \sqrt{K_a}$$

$$= 1/3 \times 18 \times 4 \times 100/18 \times\sqrt{3} - 2 \times 100 \times \sqrt{1/3}$$

$$= 230.94 - 115.47 = 115.47 \text{ kN/m}^3.$$

Question 12.14: The cohesion c, angle of internal frictionϕ and the unit weightγ of a soil are: 15kPa, 20° and 17.5kN/m³ respectively. Determine the maximum depth of unsupported excavation in the soil.

Solution:

Maximum depth of unsupported excavation,

$$H = (4c/ \; \gamma_t\sqrt{K_a})$$

$$K_a = (1 - \sin \phi)/(1 + \sin \phi)$$
$$= (1 - \sin 20°)/(1 + \sin 20°)]$$
$$= 0.49$$
$$\Rightarrow \quad H = [4 \times 15 / 17.5 \sqrt{0.49}] = 4.9 \text{ m}.$$

Question 12.15: A 3 m high vertical earth retaining wall retains a dry granular backfill with angle of internal friction of 30° and unit weight of 20 kN/m³. If the wall is prevented from yielding (no movement), determine the total horizontal thrust (in kN per unit length) on the wall.

Solution:

As well does not move, so wall is at rest.

The coefficient of earth pressure at rest

$$K_0 = 1 - \sin \phi$$
$$= 1 - \sin 30° = 0.5$$

Resultant thrust

$$P_0 = 1/2 \ (K_0 \gamma \ H)(H)$$
$$= 1/2 \times 0.5 \times 20 \times 3 \times 3$$
$$= 45 \text{ kN/m}$$

Questions

Question 1: A building is supported on a reinforced concrete raft covering an area of 14m × 21m. The subsoil is clay with an unconfined compressive strength of 84 kN/m². The pressure on the soil due to weight of the building and loads it will carry will be 120 kN/m² at the base of the raft. If the unit weight of excavated soil is 15 kN/m³, at what depth should the bottom of the raft be placed to provide a factor of safety of 3?

Question 2: A retaining wall of height 10 m consists of clay backfill. Weight of the retaining wall is 5000 kN per m acting at 3.3 m from the toe of the retaining wall. The interface friction angle between base of the retaining wall and the base soil is 20°. The depth of clay in front of the retaining wall is 2.0 m. The properties of the clay backfill and the clay placed in front of the retaining wall are the same. Assume that the tension crack is filled with water. Determine active and passive earth pressure behind the wall using Rankine's earth pressure theory. Take unit weight of water, $\gamma_w = 9.81$ kN/m³.

Question 3: The soil profile at a site consists of a 5 m thick sand layer underlain by a c - ϕ soil as shown in figure. The water table is found 1 m below the ground level. The entire soil mass is retained by a concrete retaining wall and is in the active state. The back of the wall is smooth and vertical. Determine the total active earth pressure at point A as per Rankine's theory.

Question 4: A vertical bank was formed during the excavation of a soil having $\phi = 15°$ andunit weight of 18 kN/m^3. When the depth of excavation reached 5.5 m, the bank failed. What was the approximate value of cohesion of clay?

Question 5: A vertical excavation was made in a clay deposit having unit weight of 22 kN/m^3. It caved-in after the digging reached 4m depth. Assuming $\phi = 0°$, calculate the magnitude of cohesion.

CHAPTER 13

Shallow Foundations

Question 13.1: Determine the maximum pressure which can be applied with a factor of safety of 3 through the concrete block, ensuring no bearing capacity failure in soil using Terzaghi's bearing capacity equation (without considering the shape factor, depth factor, and inclination factor). Consider width of concrete block placed at the surface as 1 m. Take $c = 0$, $N_\gamma = 40$, $\gamma_{sat} = 18$ kN/m^3; $\gamma_w = 10$ kN/m^3. Assume water table to be at the surface of the soil.

Solution: Terzaghi's bearing capacity equation for rectangular footing is given by

$$q_u = c \times N_C + \gamma \times D \times N_q + 0.5 \times \gamma' \times B \times N_\gamma$$

$$c = 0, \ B = 1, \ D = 0, \ N_\gamma = 40$$

$$\gamma_{sat} = 18 \ \text{kN/m}^3; \gamma_w = 10 \ \text{kN/m}^3$$

Since soil is submerged, $\gamma' = \gamma_{sat} - \gamma_w$

Substituting the values

$$q_u = 0 + 0 + 0.5 \times (18 - 10) \times 1 \times 40$$

$$= 160 \ \text{kN/m}^2$$

Factor of safety, (FOS) = Ultimate pressure (q_u) / actual pressure (q_a)

$$q_a = q_u/\text{FOS}$$

$$= \frac{160}{3} \times \frac{1}{2}$$

$$= 26.67 \ \text{kN/m}^2$$

$$= 26.67 \ \text{kPa.}$$

Question 13.2: The ultimate bearing capacity of a soil is 300 kN/m^2. The depth of foundation is 1m and unit weight of soil is 20 kN/m^3. Choosing a factor of safety of 2.5, determine the net safe bearing capacity.

Solution: Ultimate bearing capacity of soil, $q_u = 300 \ \text{kN/m}^2$

Depth of foundation, $D = 1$ m

Unit weight of soil, $\gamma = 20 \ \text{kN/m}^3$

Factor of safety, FOS = 2.5

Net ultimate bearing capacity of soil,

$$q_{nu} = q_u - \gamma D$$

$$= 300 - 20 \times 1 = 280 \text{ kN/m}^2$$

Net safe bearing capacity of soil,

$$q_{ns} = q_{nu} / FOS = 280/2.5 = 112 \text{ kN/m}^2.$$

Question 13.3: A square footing of 4m side is placed at 1 m depth in a sand deposit. The dry unit weight (γ) of sand is 15 kN/m³. The footing has an ultimate bearing capacity of 600 kPa. Consider the depth factors $d_q = d_\gamma = 1.0$ and the bearing capacity factor $N_\gamma = 18$. This footing is to be placed at a depth of 2m in the same sand deposit. Using Terzaghi's theory, determine the safe bearing capacity (in kPa) of this footing for a factor of safety of 3.0.

Solution:

Side of square footing = 4 m

Depth of footing = 1 m

Unit weight of soil = 15 kN/m³

Ultimate bearing capacity = 600 kPa

Depth factors, $d_q = d_\gamma = 1$

$N_\gamma = 18.75$

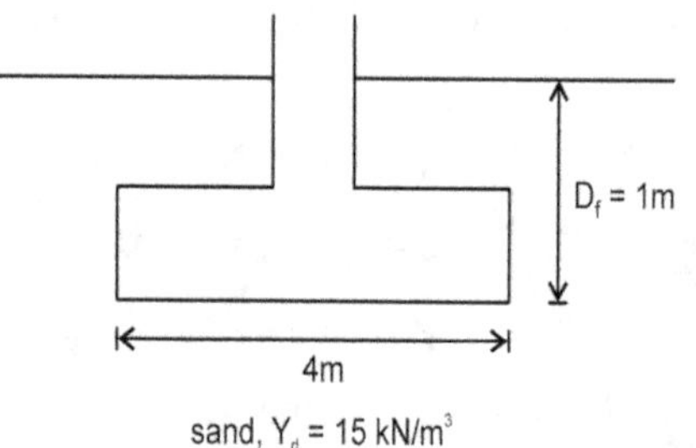

Figure 13.1

According to Terzaghi, the ultimate bearing capacity of square footing is given as

At depth of footing = 1 m

$$q_u = 1.3cN_c + qN_qd_q + 0.4\gamma BN_\gamma d_\gamma$$

For sand, c = 0, q = γD_f = 15 × 1 = 15 kN/m²

$$600 = 0 + 15 \times N_q \times 1 + 0.4 \times 4 \times 15 \times 18.75 \times 1$$

$$\Rightarrow N_q = 10$$

Now at depth of footing at 2 m,

$$q_u = 1.3\, cN_c + qN_q + 0.4B\gamma N_\gamma d_\gamma$$

$$q_u = 0 + (2 \times 15) \times 10 \times 1 + 0.4 \times 15 \times 4 \times 18.75 \times 1$$

$$q_u = 750 \text{ kPa.}$$

$\because$ We know that $q_{nu} = q_u - \gamma D_f$

$$q_{nu} = 750 - 15 \times 2$$

$$q_{nu} = 720 \text{ kPa}$$

and safe bearing capacity q_{safe}

$$q_{safe} = (q_u/FOS) + \gamma D_f$$
$$= (720/3) + 15 \times 2$$
$$q_{safe} = 270 \text{ kPa.}$$

Question 13.4: A large-scale bearing capacity test on a footing of size 1.05m x 1.05m at a depth of 1.5m yielded an ultimate value of 141 kN. Unconfined compression test on the soft saturated clay yielded strength of 0.3 N/mm². If the unit weight of the soil is 1.6 g/cm³, how much does the test value differ from that obtained using Terzaghi's bearing capacity equation?

Solution: $\gamma = 1.6 \text{g/cm}^3 = 15.596 \text{ kN/mm}^3$

$$q_u = 0.3 \text{ N/mm}^2 = 30 \text{ kN/mm}^2$$
$$c = q_u/2 = 15 \text{ kN/mm}^2$$

Taking $\phi = 0$,

We have $N_c=5.7$, $N_q=1$ and $N_\gamma=0$

For square footing, $q_f = 1.3 \, c_u \, N_c + \sigma \, N_q = 1.3 \, c_u \, N_c + \gamma D \, N_q$

For soft saturated clay we will assume soft shear failure for which

$$c = c_m = 2/3 \, c_u = 2/3 \times 15 = 10 \text{ kN/m}^2$$
$$Q_f = 1.3 \times 10 \times 5.7 + 15.96 \times 1.5 \times 1 = 97.49 \text{ kN/m}^2$$
$$Q_f = q_f \times A = 97.49 \times 1.05 \times 1.05 = 107.5 \text{ kN}$$

$Q_{test} = 141 \text{ kN}$

% difference = $(141-107.5)/107.5 = 31.2 \%$.

Question 13.5: Explain briefly how the bearing capacity is determined by field load test. The results of two field tests on a given location are as follows:

(i) Diameter = 750 mm, S = 15 mm, Q = 150 kN

(ii) Diameter = 300 mm, S = 15 mm, Q = 50 kN

Determine the load on a circular footing of 1.2 m diameter which will cause a settlement of 15 mm.

Solution: $Q = A \times q + P \times s$

$$Q_1 = 150 = \pi/4 \, (0.75)^2 \times q + \pi \times (0.75) \times s$$
$$0.4418q + 2.3562s = 150 \qquad\qquad ....(1)$$
$$Q_2 = 50 = \pi/4 \, (0.3)^2 \times q + \pi \times (0.3) \times s$$
$$0.0707q + 0.9425s = 50 \qquad\qquad ....(2)$$

Solving (1) and (2), we get

$$q = 94.22 \text{ kN/m}^2 \text{ and } s = 46 \text{ kN/m}$$

For actual footing

$$Q = 94.22 \times A + 46 \times P$$

Substituting $D = 1.2$ m

$$Q = 94.22 \times \pi/4 \times (1.2)^2 + 46 \times \pi(1.2) = 280 \text{ kN.}$$

Question 13.6: The width of a square footing and the diameter of a circular footing are equal. If both the footings are placed on the surface of a sandy soil, determine the ratio of the ultimate bearing capacity of circular footing to that of square footing.

Solution: Given

Width of square footing (B) = diameter of circular footing (D)

Depth of foundation (D_f) = 0(since both footings are placed on the surface)

For sandy soil, c = 0

Ultimate bearing capacity of circular footing

$$(q_u)_{circular} = 1.3cN_c + \gamma D_f N_q + 0.3\gamma DN_\gamma$$

Substituting c = 0 and D_f = 0

$$(q_u)_{circular} = 0.3\gamma DN_\gamma$$

Ultimate bearing capacity of square footing

$$(q_u)_{square} = 1.3cN_c + \gamma D_f N_q + 0.4\gamma BN_\gamma$$

Substituting c = 0 and D_f = 0

$$(q_u)_{square} = 0.4B\gamma N_\gamma$$

$$(q_u)_{circular}/(q_u)_{square} = 0.3D\gamma N_\gamma/0.4B\gamma N_\gamma$$

Substituting B = D; $(q_u)_{circular}/(q_u)_{square} = 0.75$.

Question 13.7: A square footing of 2 m side rests on the surface of a homogeneous soil bed having the properties: cohesion c = 24 kPa, angle of internal friction ϕ = 25° and unit weight γ = 18 kN/m³. Terzaghi's bearing capacity factors for ϕ = 25° are: N_c = 25.1; N_q = 12.7, N_γ = 9.7, N_c' = 14.8, N_q' = 5.6, N_γ' = 3.2. Determine the ultimate bearing capacity of the foundation.

Solution: For surface footing, D_f = 0

Since angle of internal friction φ is less than 29°, the soil undergoes local shear failure.

$$c_m = (2/3) \times c = (2/3) \times 24 = 16 \text{ kPa}$$

As per Terzaghi's theory

Ultimate bearing capacity for square footing is given as:

$$q_u = 1.3cN_c{}' + \gamma D_f N_q{}' + 0.4\gamma BN_\gamma{}'$$

$$q_u = 1.3 \times 16 \times 14.8 + 0 + 0.4 \times 18 \times 2 \times 3.2$$

$$q_u = 353.92 \text{ kPa.}$$

Question 13.8: A 30 cm square bearing plate settles by 8 mm in the plate load test on cohesionless soil when the intensity of loading is 180 kN/m². Determine the settlement of 1.5 m square shallow foundation under the same intensity of loading.

Solution: Given

$B_f = 1.5\text{m}$

$B_P = 0.3\text{m}$

$S_P = 8\text{mm}$

For cohesionlesss oil,

$$S_f/S_p = \{B_f(B_p+0.3)/B_p(B_f+0.3)\}^2$$

$$S_f = 8\{1.5 \times (0.3+0.3)/0.3 \times (1.5+0.3)\}^2$$

$$S_f = 22.22\text{mm.}$$

Question 13.9: Two circular footings of diameters D_1 and D_2 are resting on the surface of a purely cohesive soil. The ratio D_1/D_2 is 2. If the ultimate load carrying capacity of the footing of diameter D_1 is 200 kN/m², then determine the ultimate bearing capacity of the footing of diameter D_2.

Solution: For footings on purely cohesive soils, the bearing capacity does not depend upon the size of the footing.

Hence, bearing capacity of the footing of diameter $D_2 = 200$ kN/m².

Question 13.10: A multistorey building with a basement is to be constructed. The top 4 m of soil strata consists of loose silt, below which dense sand layer is present up to a great depth. Ground water table is at the surface. The foundation consists of the basement slab of 6 m width which will rest on the top of dense sand. For dense sand, saturated unit weight = 20 kN/m³, and bearing capacity factors are: $N_q = 40$ and $N_\gamma = 45$. For loose silt, saturated unit weight = 18kN/m³, $N_q = 15$ and $N_\gamma = 20$. Effective cohesion c' is zero for both soils. Unit weight of water is 10 kN/m³. Neglect shape factor and depth factor. Average elastic modulus E and Poisson's ratio μ of dense sand are: 60×10^3 kN/m² and 0.3 respectively.

Solution: Assume that a square footing is to be adopted and Terzaghi's analysis can be used.

From Terzaghi's formula, for c - ϕ soil

Ultimate bearing capacity $Q_f = 1.3\ c\ N_c + \sigma\ N_q + 0.4\ \gamma\ B\ N_\gamma$

Where c = cohesion, σ = effective overburden pressure, γ = unit weight of the soilunderneath B = width of base; and N_c, N_q and N_γ are bearing capacity factors.

Net ultimate bearing capacity, $Q_{nf} = Q_f - (\gamma_w \times D)$

Where, γ_w = unit weight of water, D = depth of water

Safe bearing capacity, $Q_s = Q_{nf}/FOS$; where, FOS = factor of safety.

Given:

c = 0

$\gamma_{silt} = 18\ kN/m^3$

$\gamma_{sand} = 20\ kN/m^3$

$\gamma_w = 10\ kN/m^3$

B = 6 m, D = 4 m

For sand, $N_q = 40$, $N_\gamma = 45$

For silt, $N_q = 15$, $N_\gamma = 20$

FOS = 3

$$\sigma = (\gamma_{silt} - \gamma_w) \times D$$
$$= (18 - 10) \times 4 = 32\ kN/m^2$$
$$\gamma = (\gamma_{sand} - \gamma_w)$$
$$= 20 - 10 = 10\ kN/m^3$$

Using the equation

$$Q_f = 1.3\ c\ N_c + \sigma\ N_q + 0.4\ \gamma\ B\ N_\gamma$$
$$= 0 + (32 \times 15) + (0.4 \times 10 \times 6 \times 45) = 1560\ kN/m^2$$

$\therefore Q_{nf} = 1560 - (\gamma_w \times D)$; γ_w D is to be deducted since ground water table is at ground surface.

$$= 1560 - (10 \times 4)$$
$$= 1560 - 40 = 1520\ kN/m^2$$

$\therefore$ Safe bearing capacity $Q_s = Q_{nf}/FOS = 1520/3 = 506.67\ kN/m^2$.

Question 13.11: An embankment is to be constructed by using a granular soil (bulk unit weight = 20 kN/m^3) on a saturated clayey silt deposit (undrained shear strength = 25 kPa). Assuming undrained general shear failure and bearing capacity factor of 5.7; determine the maximum height (in m) of the embankment at the point of failure.

Solution: Bulk unit weight of soil, $\gamma = 20$ kN/m^3

Undrained shear strength of soil, $c_u = 25$ kPa

Bearing capacity factor, $N_c = 5.7$

Height of embankment = H

Ultimate bearing capacity of soil, $q_u = cN_c = 25 \times 5.7 = 142.5$ kN/m^2

Maximum vertical stress due to embankment $q_u = \gamma H$

Equating ultimate bearing capacity of soil to maximum vertical stress, $q_u = \gamma H$ $142.5 = 20 \times H$

$$\Rightarrow H = 7.1 \text{ m.}$$

Question 13.12: A plate load test was conducted on sandy soil with plate of size 0.3m × 0.3m. The ultimate load per unit area was found to be 2.0 kg/cm^2. Find the allowable load per unit area for a footing 2m × 2m using a factor of safety as per Indian standard.

Solution: $Q_f = Q_p . B_f/B_p$

Or $q_f = q_p . B_f/B_p$

Therefore, $q_f = 2 \times (2/0.3)$

$$= 13.33 \text{ kg/cm}^2 = 133.33 \text{ t/m}^2$$

Using a factor of safety of 3,

$$q_a = q_f/F$$

$$= 133.33/3 = 44.4 \text{ ton/m}^2.$$

Question 13.13: The plate load test was conducted on clayey strata by using a plate of 0.3m × 0.3 m dimensions, and the ultimate load per unit area for the plate was found to be 180 kPa. Determine the ultimate bearing capacity (in kPa) of a 2 m wide square footing.

Solution: In plate load test on clayey soil,

The ultimate bearing capacity does not depend upon width of footing

Thus, $q_{uf} = q_{up} = 180$ kPa.

Question 13.14: A square footing has dimensions of 2 m × 2 m and a depth of 2 m. Determine its ultimate bearing capacity in pure clay with an unconfined strength of 0.15 N/mm^2, $\phi = 0°$ and $\gamma = 1.7$ g/cm^2. Assume Terzaghi's bearing capacity factors for $\phi = 0°$, as: $N_c = 5.7$; $N_q = 1$ and $N_\gamma = 0$.

Solution: $c = q_u/2 = 0.15/2$ N/mm²

$$= 0.075 \times 10^3 \text{ kN/m}^2$$

$$= 75 \text{ kN/m}^2$$

$$\gamma = 1.7 \text{ g/cm}^3 = 1.7 \text{ t/m}^3$$

$$= 1.7 \times 9.81 = 16.677 \text{ kN/m}^3$$

For square footing

$$q_f = 1.3\, cN_c + \sigma N_q + 0.3\, \gamma\, B\, N_\gamma$$

$$= 1.3 \times 75 \times 5.7 + 16.677 \times 2 \times 1 + 0.3 \times 1.677 \times 2 \times 0$$

$$= 589.2 \text{ kN/m}^2.$$

Question 13.15: A footing 2 m x 1 m exerts a uniform pressure of 150 kN/m² on the soil. Assuming a load dispersion of 2 vertical to 1 horizontal, determine the average vertical stress (kN/m²) at 1.0 m below the footing.

Solution: Load P $= 150 \times 2 \times 1$ kN

$$= 300 \text{ kN}$$

Area of dispersion $= (2+ 0.5+ 0.5) \times (1+ 0.5+ 0.5)$

$$= 6 \text{ m}^2$$

Vertical stress $\sigma_v = 300/6$

$$= 50 \text{ kN/m}^2.$$

Question 13.16: The observed standard penetration resistance values at different depths at a site are given below:

Depth (m)	0	0.75	1.50	2.25	3.00	3.75	4.50	5.25	6.00	6.75	7.50
Observed N value	0	5	8	10	11	13	15	17	18	22	26

Plot the observed N-values and corrected N-values at different depths. Determine the safe bearing capacity and allowable bearing pressure at a depth of 1.5 m for a square footing of 2 m width. Take unit weight of soil 18.21kN/m³.

Solution: The corrected N-values at different depths after applying overburden correction are given in table below.

The results are plotted as shown in figure 13.2.

Table 13.1 Standard penetration test record.

Depth (m)	Observed N value	Effective overburden pressure, $\sigma_0 = \gamma`H$ (kN/m²)	Overburden correction factor $C_N = 0.77 \log_{10}(2000/\sigma_0`)$	Corrected N value
0	0	0	0	0
0.75	5	6.3	1.926	9
1.50	8	12.6	1.695	13
2.25	10	18.9	1.559	15
3.00	11	25.2	1.463	16
3.75	13	31.5	1.388	18
4.50	15	37.8	1.327	19
5.25	17	44.1	1.276	21
6.00	18	50.4	1.231	22
6.75	22	56.7	1.192	26
7.50	26	63	1.156	30

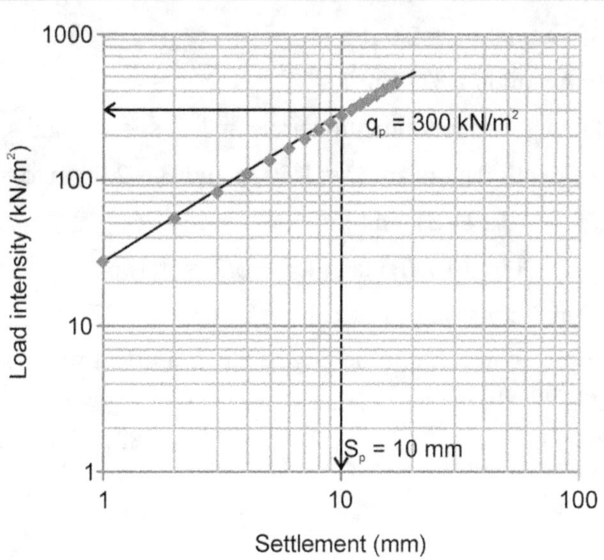

Figure 13.2: Observed and corrected standard penetration resistance values with depth.

Safe bearing capacity (shear failure criteria as per IS: 6403 - 1981):

The shear parameters may be assessed to evaluate the safe bearing capacity according to IS: 6403 -1981 for the corrected N values.

Foundation depth = 1.5 m, Foundation width = 2.0 m (square footing)

At the proposed foundation depth of 1.5 m, average N value up to a depth of twice the foundation width below depth of foundation (pressure bulb 2B, i. e., up to $2 \times 2 + 1.5 = 5.5$ m) is = $(13 + 15 + 16 + 18 + 19 + 21)/6 = 102/6 = 17$) for the site is $N_{av} = 17$. Therefore, the safe bearing capacity of the foundation soil using general shear failure is assessed below using the following equation:

$$q_u = cN_c s_c\, d_c i_c + q(N_q - 1)\, s_q d_q i_q + 0.5\, \gamma` \, BN_\gamma\, s_\gamma\, d_\gamma i_\gamma\, W` \qquad (1)$$

Where

q_u = Ultimate bearing capacity (t/m^2)

$q = \gamma` \times D$ = effective overburden pressure (t/m^2)

$\gamma`$ = submerged unit weight of soil (t/m^3)

γ = bulk unit weight of soil (t/m^3), c = cohesion $(t/m^2) = 0$.

N_c, N_q, N_γ = bearing capacity factors; s_c, s_q, s_γ = shape factors

d_c, d_q, d_γ = depth factors; i_c, i_q, i_γ = inclination factors

$W`$ = water table correction factor = 0.5,

D = depth of foundation = 1.5 m, and

B = width of foundation = 2.0 m,

Bulk density of soil is 1.821 t/m^3.

The shape factors for the foundation are: $s_c = 1.3$, $s_q = 1.2$ and $s_\gamma = 0.8$

The values of depth factors are: $D = 1.50$ m and $B = 2.00$ m, $\phi = 32.1°$

$$d_c = 1 + (0.2\, D/B)\, \tan (45° + \phi/2) = 1.272;$$

$$d_q, d_\gamma = 1 + (0.1\, D/B)\, \tan (45° + \phi/2) = 1.136$$

The inclination factors: i_c, i_q, $i_\gamma = 1$; $N_q = 24.66$, $N_\gamma = 33.16$

Considering the cohesion c to be negligible and effect of submergence, the ultimate bearing capacity, q_u

$= 0.821 \times 1.5 \times (24.66 - 1) \times 1.136 \times 1.2 \times 1 + 0.5 \times 0.821 \times 2.00 \times 33.16 \times 0.8 \times 1.136 \times 1 \times 0.5$

$= 52.09 t/m^2$.

Factor of safety = 3

Safe bearing capacity, $q_s = 52.09/3 = 17.36\ t/m^2 \approx 17.4\ t/m^2$.

Allowable bearing capacity (settlement criteria as IS: 8009 (Part-I) -1976):

As per IS: 8009 (Part-I) -1976, the safe allowable bearing (capacity) pressure has been evaluated for the permissible settlement of 50 mm as specified in IS: 1904 - 1978 for reinforced concrete structures on isolated foundations on sandy soils.

Using figure 9 of IS: 8009 (Part-I) -1976, the settlement for N_{av} = 17 for the site for a footing having size of 2 m x 2m is 16.5 mm. Therefore, the net allowable bearing (capacity) pressure for 50 mm permissible settlement, allowing water table correction (factor = 0.5) is:

$$q_n = (50 \times 10) \times 0.5/\ 16.5) = 15.15 \approx 15.2 \text{ t/m}^2.$$

Therefore, net allowable bearing capacity value for soil at site 1 = 15.2 t/m^2.

Question 13.17: The dynamic penetration resistance values at different depths at a site are given below:

Depth (m)	0.3	0.6	0.9	1.2	1.5	1.8	2.1	2.4	2.7	3.0	3.3	3.6	3.9	4.2	4.5	4.8	5.1	5.4	5.7	6.0
Dynamic cone penetration value N_{cd}	6	11	12	14	15	18	20	22	24	26	30	32	33	35	37	39	41	42	45	50

Plot the record of N_{cd} values and corresponding N-values at different depths. Determine the safe bearing capacity and allowable bearing pressure at a depth of 1.5 m for a square footing of 2 m width. Take unit weight of soil 18.21 kN/m^3.

Solution: The N-values corresponding to the dynamic cone penetration values at different depths are calculated using the correlations given below and are given in the table:

Table 13.2 Dynamic cone penetration test record.

Depth (m)	Dynamic cone penetration value N_{cd}	Corresponding standard penetration resistance value, N
0	0	0
0.3	6	4
0.6	11	7
0.9	12	8
1.2	14	9
1.5	15	10
1.8	18	12
2.1	20	13
2.4	22	14
2.7	24	16
3	26	17
3.3	30	17
3.6	32	18
3.9	33	18
4.2	35	20
4.5	37	21
4.8	39	22
5.1	41	23
5.4	42	24
5.7	45	25
6.0	50	28

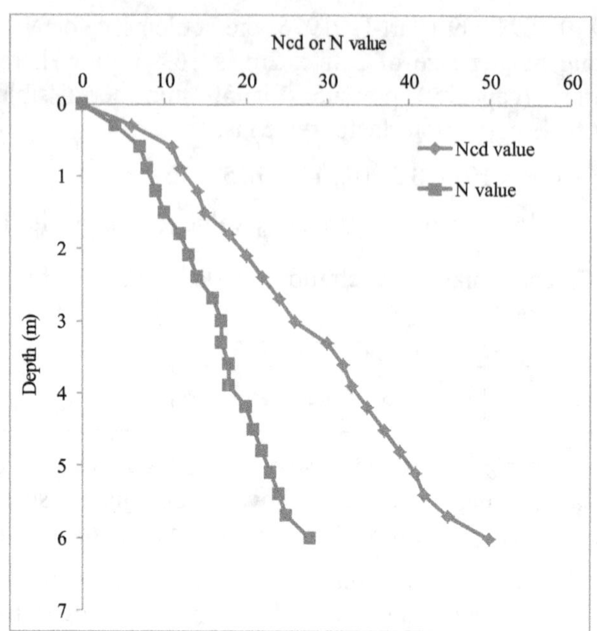

Figure 13.3: Dynamic cone penetration values and corresponding N-values with depth.

For depth, D up to 3 m; $N = N_{cd}/1.50$, (i.e. $6/1.50 = 4$).

For depth, $3 < D \leq 6$ m; $N = N_{cd}/1.75$, (i.e. $30/1.75 = 17$).

For depth, $D > 6$ m, $N = N_{cd}/2.0$.

The results are plotted as shown in figure 13.3.

Safe bearing capacity (shear failure criteria as per IS: 6403 - 1981):

The shear parameters may be assessed to evaluate the safe bearing capacity according to IS: 6403 -1981 for the corrected N values.

Proposed foundation depth = 1.5 m,

Proposed foundation width = 2.0 m (square footing)

At the proposed foundation depth of 1.5 m, average N value up to a depth of twice the foundation width below depth of foundation (pressure bulb 2B, i. e., up to $2 \times 2 + 1.5 = 5.5$ m) is = $(10 + 12 + 13 + 14 + 16 + 17 + 17 + 18 + 18 + 20 + 21 + 22 + 23 + 24)/14 = 245/14 = 17$) for site 1 is $N_{av} = 17$. Therefore, the safe bearing capacity of the foundation soil for general shear failure is assessed below using the following equation:

$$q_u = cN_c s_c\, d_c i_c + q(N_q -1)\, s_q d_q i_q + 0.5\, \gamma\, BN_\gamma\, s_\gamma\, d_\gamma i_\gamma\, W` \qquad (1)$$

Where

q_u = Ultimate bearing capacity (t/m^2)

$q = \gamma` \times D$ = effective overburden pressure (t/m^2)

γ' = submerged unit weight of soil (t/m^3)

γ = bulk unit weight of soil (t/m^3), c = cohesion (t/m^2) = 0.

N_c, N_q, N_γ= bearing capacity factors; s_c, s_q, s_γ = shape factors

d_c, d_q, d_γ = depth factors; i_c, i_q, i_γ = inclination factors

W` = water table correction factor = 0.5,

D = depth of foundation =1.5 m, and

B = width of foundation = 2.0 m,

Bulk density of soil is 1.821 t/m^3.

The shape factors for the foundation are:s_c = 1.3, s_q = 1.2 and s_γ = 0.8

The values of depth factors are:D = 1.50 m and B = 2.00 m, ϕ = 32.1°

$$d_c = 1+ (0.2\ D/B)\ \tan (45° + \phi/2) = 1.272;$$

$$d_q,\ d_\gamma = 1+ (0.1\ D/B)\ \tan (45° + \phi/2) = 1.136$$

The inclination factors: i_c, i_q, i_γ = 1; N_q = 24.66, N_γ = 33.16

Considering the cohesion c to be negligible and effect of submergence, the ultimate bearing capacity,q_u

= 0.821 × 1.5 ×(24.66 − 1)× 1.136 × 1.2 × 1 + 0.5 × 0.821× 2.00 × 33.16 × 0.8× 1.136 × 1 × 0.5

= 52.09 t/m^2.

Factor of safety = 3

Safe bearing capacity, q_s= 52.09/3 = 17.36 t/m$^2 \approx$ 17.4 t/m^2.

Allowable bearing capacity (settlement criteria as per IS:8009 (Part-I) - 1976):

As per IS: 8009 (Part-I) -1976, the safe allowable bearing (capacity) pressure has been evaluated for the permissible settlement of 50 mm as specified in IS: 1904 - 1978 for reinforced concrete structures on isolated foundations on sandy soils.

Using figure 9 of IS: 8009 (Part-I) -1976, the settlement for N_{av} = 17 for the site for a footing having size of 2 m x 2m is 16.5 mm. Therefore, the net allowable bearing (capacity) pressure for 50 mm permissible settlement, allowing water table correction (factor = 0.5) is:

$$q_n= (50 × 10) × 0.5/\ 16.5)$$

$$= 15.15$$

$$\approx 15.2\ t/m^2.$$

Therefore, net allowable bearing capacity value for soil at site 1 = 15.2 t/m^2.

Question 13.18: The load and corresponding penetration dial readings in a plate load test conducted on 300 mm square plate are given below:

Load (kN)	0	2.5	5.0	7.5	10	12.5	15	17.5	20	22.5	25	27.5	30	32.5	35	37.5	40	42.5
Penetration dial reading	0	113	218	340	455	587	696	872	1055	1238	1385	1650	1883	2217	2450	2728	2989	3304

The least count of dial is:

1 small division = 0.01 mm.

Plot the load intensity – settlement curves using a simple scale graph and and log-log graph. Determine the safe bearing capacity and settlement of 1 m wide footing over clayey soil.

Solution: Load intensity = load/area

$$= 2.5/0.09$$

$$= 27.78 \text{ kN/m}^2.$$

Settlement = Penetration dial reading $\times$ least count

$$= 113 \times 0.01$$

$$= 1.13 \text{ mm.}$$

The load intensity and the corresponding settlement values are given in the table below.

Load intensity-settlement calculations

Load (kN)	Penetration dial reading	Load intensity = Load/area (kN/m^2)	Settlement (mm)
0	0	0	0
2.5	113	27.78	1.13
5.0	218	55.56	2.18
7.5	340	83.33	3.40
10.0	455	111.11	4.55
12.5	587	138.89	5.87
15.0	696	166.67	6.96
17.5	872	194.44	8.72
20.0	1055	222.22	10.55
22.5	1238	250.00	12.38
25.0	1385	277.78	13.85
27.5	1650	305.56	16.50
30.0	1883	333.33	18.83
32.5	2217	361.11	22.17
35.0	2450	388.89	24.50
37.5	2728	416.67	27.28
40.0	2989	444.44	29.89
42.5	3304	472.22	33.04

The load intensity - settlement curves are plotted on simple scale graph and and log-log graph as shown in figures 13.4 and 13.5 respectively.

From figures 13.4 and 13.5, the ultimate bearing capacity of plate, $q_p = 300$ kN/m^2.

Settlement of plate, $S_p = 10$ mm.

For clayey soil, ultimate bearing capacity of footing, $q_f \approx q_p = 300$ kN/m^2.

Safe bearing capacity of footing, $q_s = 300/3 = 100$ kN/m^2.

Factor of safety $= 3$.

The settlement S of the footing of width B can be calculated from the settlement of the plate S_p using the following expression:

$$S = S_p \, B/ \, B_p$$

$$= 10 \times 1/0.3 = 33.33 \text{ mm}$$

Safe bearing capacity of footing, $q_s = 100$ kN/m^2.

Settlement of the footing $S = 33.33$ mm.

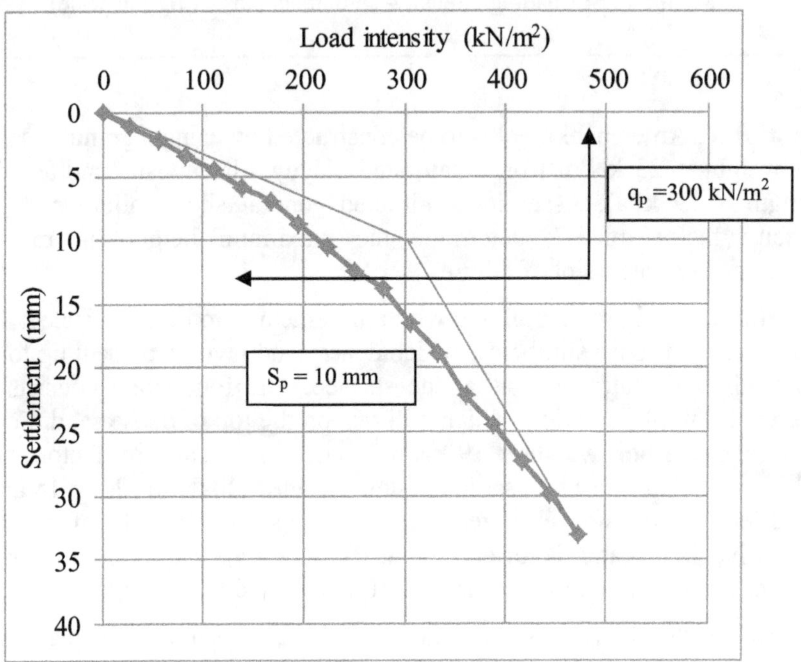

Figure 13.4: Loading intensity- settlement curve.

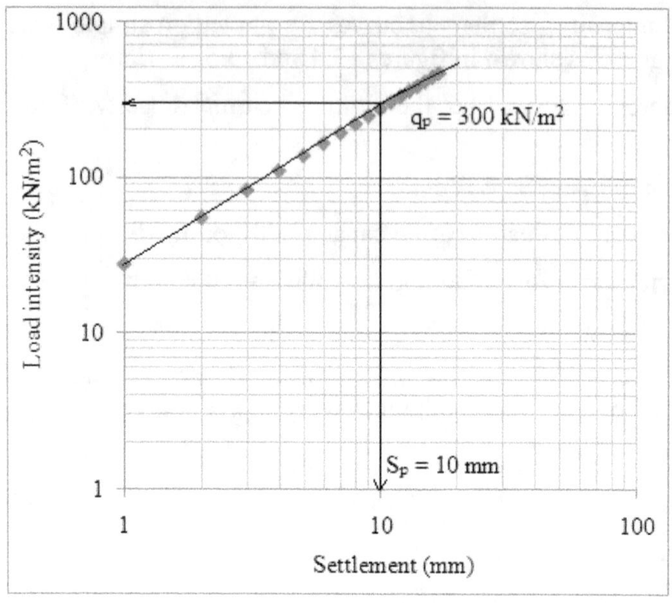

Figure 13.5: Loading intensity- settlement curve (log-log scale).

 **Questions**

Question 1: An embankment is to be constructed by using a granular soil (bulk unit weight = 23 kN/m^3) on a saturated clayey silt deposit (undrained shear strength = 19 kPa). Assuming undrained general shear failure and bearing capacity factor of 5.7, determine the maximum height (in m) of the embankment at the point of failure.

Question 2: A high-rise building with a basement is to be constructed. The top 3 m consists of loose silt, below which dense sand layer is present up to a great depth. Ground water table is at the surface. The foundation consists of the basement slab of 7 m width which will rest on the top of dense sand. For dense sand, saturated unit weight = 19 kN/m^3, and bearing capacity factors N_q = 40 and N_γ = 45. For loose silt, saturated unit weight = 20kN/m^3, N_q = 15 and N_γ = 20. Effective cohesion c' is zero for both soils. Unit weight of water is 10 kN/m^3. Neglect shape factor and depth factor. Average elastic modulus E and Poisson's ratio μ of dense sand is 50 x 10^{3}kN/m^2 and 0.35 respectively.

Question 3: A 60 cm square bearing plate settles by 8 mm in the plate load test on cohesionless soil when the intensity of loading is 200 kN/m^2. Determine the settlement of a shallow foundation of 2.5 m square under the same intensity of loading.

Question 4: A bearing capacity test on a footing of size 2 m x 2 m at a depth of 1.5m yielded an ultimate value of 170 kN. Unconfined compression test on the soft saturated clay yielded strength of 0.4 N/mm^2. If the unit weight of the soil is 1.7 g/cc, how much does the test value differ from that obtained using Terzaghi's bearing capacity equation?

Question 5: The ultimate bearing capacity of a soil is 348 kN/m^2. The depth of foundation is 1.5m and unit weight of soil is 21 kN/m^3. Choosing a factor of safety of 2.5, determine the net safe bearing capacity.

Question 6: A building is to be supported on a R.C. raft covering an area of 14m × 21m. The subsoil is clay, with an unconfined compressive strength of 84 kN/m^2. The pressure on the soil due to the weight of the building and the loads that it will carry, will be 120 kN/m^2 at the base of the raft. If the unit weight of excavated soil is 15kN/m^3, at what depth should the bottom of the raft be placed to provide a factor of safety of 3?

Question 7: Determine the percent reduction in the bearing capacity of a strip footing resting on sand under flooding condition (water level at the base of the footing) when compared to the situation where the water level is at a depth much greater than the width of footing.

Question 8: The width of a square footing is 1.5 times the diameter of a circular footing and footings are placed on the surface of sandy soil. Determine the ratio of the ultimate bearing capacity of circular footing to that of square footing.

Question 9: An embankment is to be constructed by using granular soil (bulk unit weight = 20 kN/m^3) on a saturated clayey silt deposit (undrained shear strength = 25 kPa). Assuming undrained general shear failure and bearing capacity factor of 5.7, determine the maximum height (in m) of the embankment at the point of failure.

Question 10: A building has to be supported over R.C. raft foundation of dimensions 14m × 21m. The soil is clay, which has an average unconfined compressive strength of 15 kN/m^2. The pressure on the soil due to the weight of the building and the loads that it will carry, will be 140 kN/m^2 at the base of the raft. The building has provision for basement floors. At what depth should the bottom of the raft be placed to provide a factor of safety of 3 against shear failure? γ_{clay} = 19kN/m^3. Use Skemptons approach for bearing capacity calculations.

CHAPTER 14

Deep Foundations

Question 14.1: A 0.5 m × 0.5 m square concrete pile is to be driven in a homogeneous clayey soil having with undrained shear strength, $c_u = 50$ kPa and unit weight, $\gamma = 18.0$ kN/m^3 (figure 14.1). The design capacity of the pile is 500 kN. The adhesion factor α is given as 0.75. Determine the length of the pile required for the above design load with a factor of safety of 2.0.

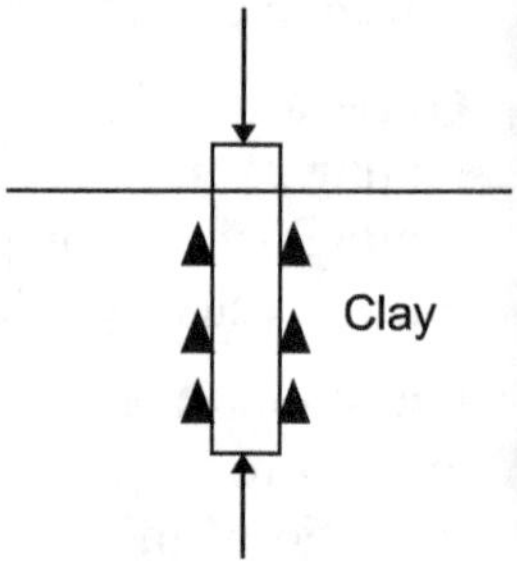

Solution: Given, size of pile = 0.5 m × 0.5 m

Undrained shear strength, $c_u = 50$ kPa

Unit weight, $\gamma = 18$ kN/m^3

Adhesion factor, $\alpha = 0.75$, $N_c = 9$

Factor of safety = 2

Design capacity of pile = 500 kN

∴ Ultimate capacity of pile $Q_u = 500 \times 2 = 1000$ kN

$$Q_u = cN_cA_P + \alpha C_uPL$$

$$1000 = 9 \times 50 \times 0.5 \times 0.5 + 0.75 \times 50 \times 4 \times 0.5 \, L$$

$$L = 11.83 \text{ m or } L \simeq 11.8 \text{ m.}$$

Figure 14.1

Question 14.2: A group of 9 piles, 10m long is used as a foundation for bridge pier. The piles used are 30cm diameter centre to centre spacing of 0.9m. The subsoil consists of clay with unconfined compressive strength of 1.5 kg/cm^2. Determine the efficiency neglecting the bearing action. Given: adhesion factor = 0.9.

Solution: B = 2 × 0.9 + 0.3 = 2.1m

$$c = q_u/2 = 1.5/2 = 0.75 \text{kg/cm}^2 = 7.5 \text{ t/m}^2$$

(a) For piles acting individually

$$Q_{un} = n. \, m. \, c. \, A_s$$

$$= 9 \times 0.9 \times 7.5(\pi \times 0.3 \times 10) = 572.6 t$$

(b) Piles acting in a group

$$Q_{ug} = c \times (4BL) = 7.5 \times 4 \times 2.1 \times 10 = 630 t$$

Efficiency for pile group, $\eta = Q_{ug}/Q_{un} = 630/572.6 = 1.1 t$

Question 14.3: A single vertical friction pile of diameter 500 mm and length 20 m is subjected to vertical compressive load. The pile is embedded in a homogeneous sandy stratum where: angle of internal friction (φ) = 30°, dry unit weight (γ_d) = 20 kN/m^3 and angle of wall friction (δ) = $2\varphi/3$. Considering the coefficient of lateral earth pressure K = 2.7 and the bearing capacity factor Nq = 25, determine the ultimate bearing capacity of the pile.

Solution: Ultimate bearing capacity of a pile is given by

$$Q_s = Q_b + Q_s$$

Where,

Q_b= end bearing resistance = $A_b \, f_b$

Q_s= skin frictional resistance or shaft resistance =$A_s f_s$

For sandy soils:

$$Q_u = A_b \sigma'_v N_q + A_s \, K\sigma'_{vavg} \, \tan\delta$$

σ'_v= effective vertical stress at pile level

N_q= depends on φ values

K= coefficient of lateral earth pressure

σ'_{vavg}= effective vertical average stress along the pile length

δ = angle of friction between pile and soil

Given

$$D = 500mm, \; L=20m, \; \varphi=30°$$

$$\gamma_d = 20kN/m^3,$$

$$\delta = 2/3\varphi,$$

$$K=2.7, \; N_q=25$$

For friction pile for sandy soils

$$Q_u = A_s \, f_s$$

$$f_s =1/2\sigma_v K\tan\delta$$

$$\sigma_v = \gamma_d \times L= 20 \times 20 =400 \; kN/m^2$$

$$\tan\delta = \tan(2/3\varphi) =0.364$$

$$f_s =1/2 \times 400 \times 2.7 \times 0.364=196.56 kN/m^2$$

$$Q_u =196.56 \times \pi dL$$

$$Q_u =196.56 \times \pi \times 0.5 \times 20=6175kN$$

The ultimate bearing capacity= 6175kN

Question 14.4: A timber pile of length 8 m and diameter 0.2 m is driven with a 20 kN drop hammer, falling freely from a height of 1.5 m. The total penetration of the pile in the last 5 blows is 40 mm. Use Engineering News-Record formulas. Assume a factor of safety of 6 and an empirical factor (allowing reduction in the theoretical set, due to energy losses) of 2.5 cm. Determine the safe load-carrying capacity of the pile.

Solution: Length of the pile, $L = 8$ m

Diameter of the pile, $D = 0.2$ m

Weight of drop hammer, $W = 20$ kN

Height of falling of drop hammer, $h = 1.5$ m $= 150$ cm

Total penetration of the pile in last 5 blows $= 40$ mm

Set value, $S = 8$ mm $= 0.8$ cm

Factor of safety, $F = 6$

Empirical factor, $c = 2.5$cm

Safe load carrying capacity of pile, $Q_s = ?$

According to Engineering News Record formula,

$$Qs = \frac{w \times h \times n_h}{F(s + c)} = \frac{20 \times 150 \times 1}{6(0.8 + 2.5)} = 151.25 \text{ kN.}$$

Question 14.5: Two identical concrete piles having the plan dimensions 50cm×50cm each are driven into a homogeneous sandy layer as shown in the figure. Consider the bearing capacity factor N_q for $30°$ as 24. If Q_{P1} and Q_{P2} represent the ultimate point bearing resistances of the piles under dry and submerged conditions respectively, determine the ratio Qp_1/Qp_2.

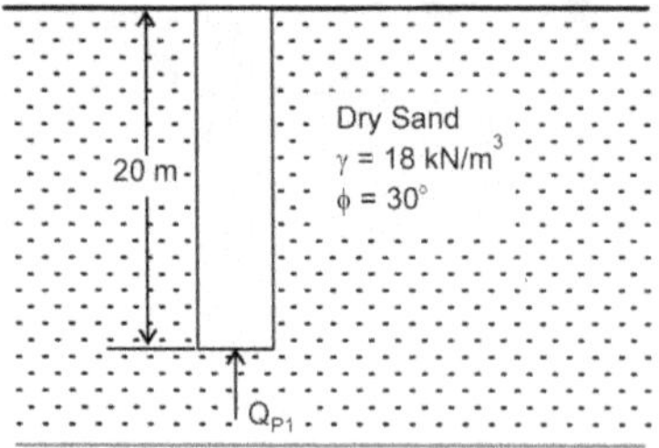

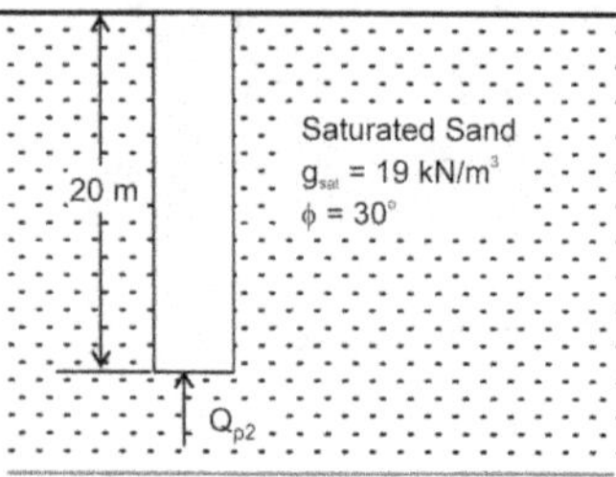

Figure 14.2

Solution: $Q_{P1} = qN_q + 0.4\gamma BN_\gamma$

$$= 18 \times 20 \times N_q + 0.4 \times 0.5 \times 18 \times N_\gamma$$

$$= 18 \,(20N_q + 0.2N_\gamma)$$

$$Q_{P2} = qN_q + 0.4\gamma BN_\gamma$$

$$= (19 - 10) \times 20 \times N_q + 0.4 \times 0.5 \times (19 - 10)\, N_\gamma$$

(Assuming $\gamma_w = 10 \text{kN/m}^3$)

$$Q_{P2} = 9 \ (20N_q + 0.2N_\gamma)$$

$$Q_{p1} / Q_{p2} = 2$$

Hence, $Q_{p1} > Q_{p2}$ by about 100%

Question 14.6: The ultimate load capacity of a 10 m long concrete pile of square cross section 500 mm × 500 mm driven into a homogenous clay layer having undrained cohesion value of 40 kPa is 700 kN. If the cross section of the pile is reduced to 250 mm × 250 mm and the length of the pile is increased to 20 m, determine the ultimate load capacity.

Solution: For pile 1:

Ultimate load carrying capacity is given by

$$Q_u = c_u N_c A_b + m c_u P_b L$$

Ultimate load carrying capacity, $Q_u = 700$ kN

Cohesion, $c_u = 40$ kPa

Bearing capacity factor, $N_c = 9$

Bearing area, $A_b = 0.5 \times 0.5$

Adhesion factor, $m = ?$

Perimeter of the pile,

$$P_b = 4 \times 0.5$$

Length of the pile, L = 10 m

$$700 = 40 \times 9 \times 0.5 \times 0.5 + m \times 40 \times 4 \times 0.5 \times 10$$

m = 0.7625

For pile 2:

$$c_u = 40 \text{ kPa},$$

$N_c = 9,$

$$A_b = 0.25 \times 0.25$$

$$m = 0.7625,$$

$$P_b = 4 \times 0.25,$$

$$L = 20 \text{ m}$$

Ultimate load carrying capacity is given by

$$Q_u = c_u N_c A_b + m c_u P_b L$$

$$Q_u = 40 \times 9 \times 0.25 \times 0.25 + 0.7625 \times 40 \times 4 \times 0.25 \times 20$$

$$= 632.5 \text{ kN}$$

Question 14.7: Estimate the safe load carrying capacity of a single bored pile 20m long and 500mm diameter. The adhesion coefficient α is 0.4 and take factor of safety as 2.5. The details of soil strata are as follows:

Depth(m)	Soil deposit	Undrained shear strength (kPa)
0 - 5	Loose soil	50
5-10	Weathered over consolidated clay	70
10-15	Over consolidated clay	100
15-20	Highly over consolidated clay	200

Assume that $\varphi_u = 0$ is valid and $N_c = 9$ for deep foundation.

Solution: $Q_{up} = A_p r_p + A_s r_f$

Where A_p = point bearing area = $3.14 \times (0.5 \times 0.5) = 0.1963$ m^2

$A_s = 3.14 \times d \times L = 3.14 \times 0.5 \times 20 = 31.416$ m^2

$r_p = cN_c = 9c_u = 9s_u = 9 \times 200 = 1800$ kN/m^2

$r_f = \alpha c_{ua} = 0.4 c_{ua}$

Where c_{ua} = average shear strength of soil over the depth of pile

$c_{ua} = (50 \times 5 + 70 \times 5 + 100 \times 5 + 200 \times 5) / 20 = 105$ kN/m^2

$Q_{up} = 0.1963 \times 1800 + 31.416 \times 0.4 \times 105 = 1672.8$ kN

Safe load, $Q_s = Q_{up}/F = 1672.8/2.5 = 669.1$ kN.

Question 14.8: A group of nine piles in a square pattern is embedded in soil strata comprising dense sand underlying the recently filled clay layer, as shown in the figure. The perimeter of an individual pile is 126 cm. The size of pile group is 240 cm × 240 cm. The recently filled clay has undrained shear strength of 15 kPa and unit weight 16 kN/m^3. Determine the negative frictional load acting on pile group.

Solution: The perimeter of individual pile, P = 126 cm = 1.26 m

Size of pile group = 2.40 m × 2.40 m

Undrained shear strength of recently filled clay,

c = 15 kPa

Unit weight of recently filled clay, = 16 kN/m^3

Negative friction load acting on the pile group = ?

For individual action of piles, negative skin friction,

$$F_{ng} = nF_n = n.\,\alpha.\,c.\,p.\,L$$

$$= 9 \times 1 \times 15 \times 1.26 \times 2 = 340.2 \text{ kN}$$

For group action of piles, negative skin friction

F_{ng} = n. α. c. p. L + weight of soil in negative zone

$$= (1\times 15\times 4\times 2.4\times 2)+ (2.4\times 2.4\times 2\times 16)$$

$$= 288+184.32 = 472.32\text{kN}$$

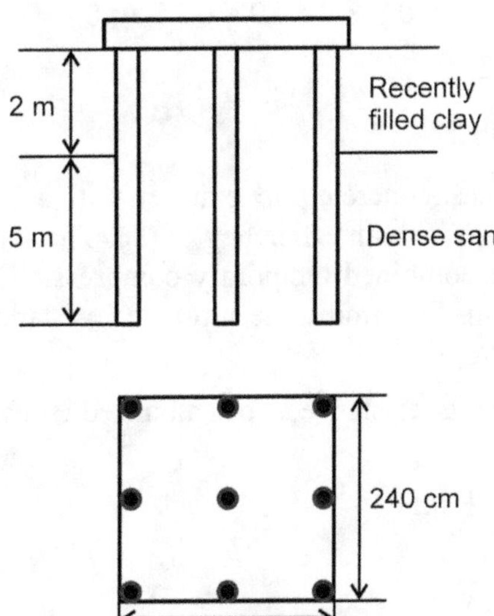

Figure 14.3

Negative skin friction is the maximum value of above two cases = 472.32 kN.

Question 14.9: A 0.5 m × 0.5 m square concrete pile is to be driven in a homogeneous clayey soil having undrained shear strength, = 50 kPa and unit weight, =18.0 kN/m^3. The design capacity of the pile is 500 kN. The adhesion factor is given as 0.75. Determine the length of the pile required for the above design load with a factor of safety of 2.0.

Solution: Size of concrete pile = 0.5 m×0.5 m

Undrained shear strength of soil, c_u = 50 kPa

Unit weight of soil, = 18 kN/m^3

Design capacity of pile, = 500 kN

Adhesion factor, = 0.75

Factor of safety, F = 2.0

Length of the pile, L = ?

For clayey soil, φ = 0

$N_c = 9$

$$f_b = cN_c = 9\,c_u$$

$$Q_u = Q_b + Q_s$$

$$FQ_u = A_b f_b + \alpha c_u\,A_s$$

$$FQ_u = A_b 9 c_u + \alpha c_u\,A_s$$

$$2 \times 500 = 0.5 \times 0.5 \times 9 \times 50 + 0.75 \times 50 \times 4 \times 0.5 \times L$$

$$1000 = 112.5 + 75L$$

$$L = 11.83\text{m}$$

Question 14.10: A precast concrete pile is driven with a 50 kN hammer railing through a height of 1.0 m with an efficiency of 0.6. The set value observed is 4 mm per below and the combined temporary compression of the pile, cushion and the ground is 6 mm. Determine the ultimate resistance of the pile as per Modified Hiley Formula.

Solution: Ultimate driving resistance as per modified Hiley's formula is given by

$$R = Wh\eta\,/(S + 0.5C)$$

$W = 50$ kN

$h = 1$ m

$\eta = 0.6$

$S = 4$ mm $= 4 \times 10^{-3}$m

$C = 6$ mm $= 6 \times 10^{-3}$ m

Substituting the values

$$R = 50 \times 1 \times 0.6/(4 \times 10^{-3} + 0.5 \times 6 \times 10^{-3})$$

$$= 4285.7\text{kN}.$$

Question 14.11: From the load – settlement data for a micro pile of 50 mm diameter, plot the load-settlement curve for the pile. Determine the ultimate load corresponding to a settlement of 10% of pile diameter.

Load (kN)	0	10	20	30	40	50	60	70	80	90	100	110	120
Settlement (mm)	0	0.10	0.18	0.30	0.45	0.80	1.40	2.15	3.10	4.20	5.90	8.50	15.10

Solution: The load – settlement curve for the micro pile is plotted as shown in the figure 14.4.

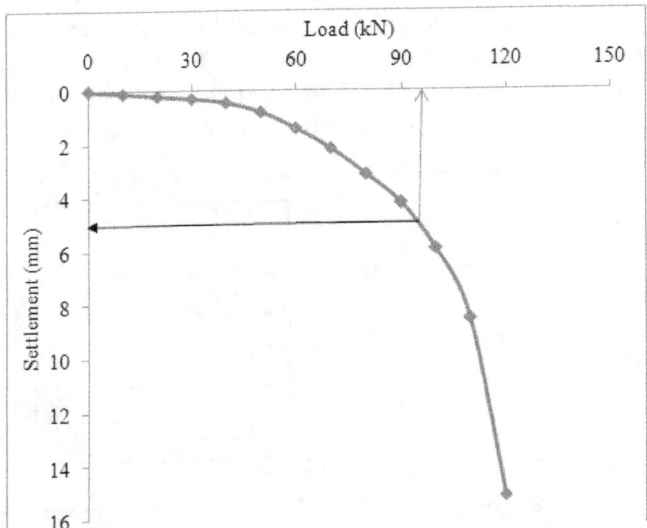

Figure 14.4: Load - settlement curve for micro pile.

From figure 14.4, the ultimate load corresponding to a settlement of 10% of the pile diameter

(i.e., 5 mm) = 95kN.

Questions

Question 1: A group of 16 piles of 10m length and 0.5m diameter is installed in 10m thick stiff clay layer underlain by rock. The pile soil adhesion factor is 0.4, average shear strength on both sides is 100 kPa, and undrained shear strength of soil at base is also 100kPa. Determine the base resistance of single pile.

Question 2: A group of nine piles in a 3 × 3 square pattern is embedded in a soil stratum comprising dense sand underlying a recently filled clay layer. The perimeter of an individual pile is 126 cm. The size of the pile group is 240 cm × 240 cm. The recently filled clay has undrained shear strength of 15 kPa and unit weight of 16kN/m^3. Determine the negative frictional load acting on the pile group.

Question 3: A building is to be supported on reinforced concrete raft covering area of 14 m× 21 m. The subsoil is clay with an unconfined compressive strength of 84 kN/m^2. The pressure on the soil due to weight of the building and loads it will carry will be 120 kN/m^2 at the base of the raft. If the unit weight of excavated soil is 15 kN/m^3, at what depth should the bottom of the raft be placed to provide a factor of safety of 3?

Question 4: A group of 16 piles was installed in a layered clay soil deposit as shown in figure 14.5. The diameter of piles is 500 mm and their c/c distance is 1m. The length of the pile group is 18m. Estimate the safe load capacity of the group with a factor of safety of 2.5. The adhesion factor between the pile and soil in each soil layer are shown in the figure below.

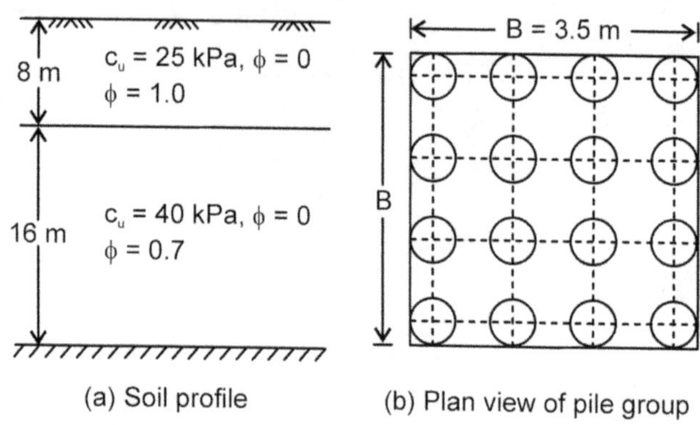

(a) Soil profile   (b) Plan view of pile group

Figure 14.5

Question 5: A square concrete pile is to be driven in a homogeneous clayey soil having undrained shear strength 100 kPa and unit weight23.0 kN/m^3. The design capacity of the pile is 600 kN. The adhesion factor is given as 0.79. Determine the length of the pile required for the above design load with a factor of safety of 1.8.

Question 6: A timber pile of length 6 m and diameter 0.4 m is driven with a 25 kN drop hammerfalling freely from a height of 1.5 m. The total penetration of the pile in the last 5 blows is 40 mm. Use the Engineering News Record expression. Assume a factor of safety of 6and empirical factor (allowing reduction in the theoretical set, due to energy losses) of2.5 cm. Determine the safe load carrying capacity of the pile.

Question 7: A single vertical friction pile of diameter 500 mm and length 20 m is subjected to vertical compressive load. The pile is embedded in a homogeneous sandy stratum where: angle of internal friction, φ = 30°, dry unit weight, γ_d = 20 kN/m^3 and the angle of wall friction, δ = 2φ/3. Considering the coefficient of lateral earth pressure, K = 2.7 and the bearing capacity factor, N_q= 25, determine the ultimate bearing capacity of the pile.

CHAPTER 15

Soil Exploration

Question 15.1: 75 mm is the external diameter of a sampling tube. If the area ratio required is 20%, determine the thickness of the sampling tube. In what type of clay would such a high area ratio be required?

Solution:

$$A_r = (D_o{}^2 - D_i{}^2)/D_i{}^2$$

$$0.20 = (75^2 - D_i{}^2)/D_i{}^2$$

$$0.20 D_i{}^2 + D_i{}^2 = 5625$$

$$1.2 D_i{}^2 = 5625$$

$$D_i{}^2 = \frac{5625}{1.2} = 4687.5$$

$$D_i = \sqrt{(4687.5)} = 68.465 \text{ mm}$$

The wall thickness $= \dfrac{75 - 68.465}{2} = 3.26$ mm

When samples are to be taken in very stiff to hard clay soils mixed with stones, sampling tube with high area ratio are required.

Question 15.2: During a sampling operation, the driver sampler is advanced 600 mm and the length of the sample recovered is 525 mm. What is the recovery ratio of sample?

Solution:

Recovery ratio is defined as

$$\text{Recovery ratio (R)} = \frac{L}{H} \times 100\%$$

$$R = \frac{525}{600} = 0.875 < 1$$

If $R < 1$, this indicates that the sample is compressed.

Question 15.3: A large number of undisturbed samples are to be obtained using 100 mm inside diameter of sampling tube. Making suitable assumptions, determine the thickness of the sampling tube.

Solution:

$D_i = 100$ mm

Assuming area ratio = 20%

$$A_r = (D_o^2 - D_i^2)/D_i^2$$
$$0.20 = (D_o^2 - (100)^2)/(100)^2$$

Outer diameter $D_o = 109.54$ mm

$$\text{Wall thickness} = \frac{109.54 - 100}{2} = 4.77 \text{ mm.}$$

Question 15.4: The inner diameter of sampling tube and that of cutting edge are 70 mm and 68 mm respectively. Their outer diameters are 72 mm and 74 mm respectively. Find the inside clearance, outside clearance and area ratio of the sampler.

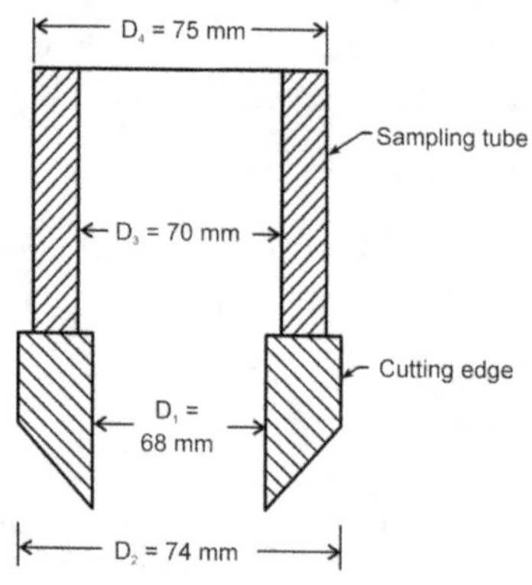

Figure 15.1

Solution:

1. Inside clearance, $C_i = \dfrac{D3 - D1}{D1} \times 100$

$$= \frac{70 - 68}{68} \times 100 = 2.941\%$$

2. Outside clearance, $C_o = \dfrac{D2 - D4}{D4} \times 100$

$$= \frac{74 - 72}{72} \times 100 = 2.777\%$$

3. Area ratio, $A_r = (D_2^2 - D_1^2)/D_1^2 \times 100$

$$= [\{(74)^2 - (68)^2\}/(68)^2] \times 100 = 18.42\%$$

Question 15.5: The most common soil and soft rock sampling tool in the US is the Standard split spoon. Split spoon tubes split longitudinally into halves and permit taking a soil or soft rock sample. The tube size is designated as an NX. The NX outside diameter is $D_o = 50.8$ mm (2 inches) and its inside diameter is $D_i = 34.9$ mm (1-3/8 inches). This small size has the advantage of cheapness, because it is relatively easy to drive into the ground. However, it has the disadvantage of disturbing the natural texture of the soil. In soft rocks, such as young limestone, it will destroy the rock to such a degree that it may be classified as "sand".

A better sampler is the Shelby (or thin-tube sampler). It has the same outside diameter of 2 inches (although the trend it to use 3 inches).

Compare the degree of sample disturbance of a US standard split-spoon sampler, versus the two Shelby thin-tube samplers (2" and 3"outside diameters) via their area ratio A_r (a measure of sample disturbance).

Solution:

The area ratio for a 2" standard split spoon sampler is

$$A_r = (D_2{}^2 - D_1{}^2)/D_1{}^2 = ((2)^2 - (1.38)^2)/(1.38)^2 = 1.1 = 110\%$$

The area ratio for a 2" Shelby-tube sampler is

$$A_r = (D_2{}^2 - D_1{}^2)/D_1{}^2 = ((2)^2 - (1.875)^2)/(1.875)^2 = 0.138 = 13.8\%$$

The area ratio for a 3" Shelby-tube sampler is

$$A_r = (D_2{}^2 - D_1{}^2)/D_1{}^2 = ((3)^2 - (2.875)^2)/(2.875)^2 = 0.089 = 8.9\%$$

Clearly, the 3"(76 mm) outside diameter Shelby-tube sampler is the best tool to use.

Question 15.6: During a sample operation, the drive sampler is advanced 600mm and the length of the sample recovered is 525mm. What is the recovery ratio of the sample?

Solution:

Recovery ratio is defined as

$$R_r = \frac{\text{Actual length of sample in the tube}}{\text{Total length of the sampling tube driven below the bottom of the bore hole}}$$

$$R_r = 525/600 = 0.875 < 1$$

Since $R_r < 1$, this indicates that the sample is compressed.

Question 15.7: A soil sampler has inner and outer radii of 25mm and 30mm, respectively. Determine the area ratio of the sampler.

Solution:

$$\text{Area ratio} = \frac{D_0{}^2 - D_i{}^2}{D_i{}^2} \times 100 = \frac{30^2 - 25^2}{25^2} \times 100 = 44\%.$$

Question 15.8: The following sizes of sampling tubes are available in the market:

Sampling tube 1: O.D. = 70 mm, I.D. = 67 mm, length = 600 mm

Sampling tube 2: O.D. = 120 mm, I.D. = 117 mm, length = 600 mm

Sampling tube 3: O.D. = 55 mm, I.D. = 47 mm, length = 600 mm

Out of these which one would you select for obtaining undisturbed soil samples from a borehole, if the diameter of hole is 100mm?

Solution:

Area ratio for sampling tube 1:

$$A_r = \frac{D_0{}^2 - D_i{}^2}{D_i{}^2} \times 100 = \frac{70^2 - 67^2}{67^2} = 9.15\ \%$$

Area ratio for sampling tube 2:

$$A_r = \frac{D_0{}^2 - D_i{}^2}{D_i{}^2} \times 100 = \frac{120^2 - 117^2}{117^2} = 5.19\%$$

Area ratio for sampling tube 3:

$$A_r = \frac{D_0{}^2 - D_i{}^2}{D_i{}^2} \times 100 = \frac{55^2 - 47^2}{47^2} = 36.94\%$$

For obtaining the soil samples as undisturbed, the area ratio should be less than 10% which is for sampling tube 1 and sampling tube 2.

Since the diameter of the borehole is 100mm, we cannot take sampling tube with diameter more than the diameter of the borehole as it will disturb the soil in vicinity.

Therefore, sampling tube 1 is used to obtain undisturbed soil sample.

Question 15.9: A sampling tube has inner diameter of 70 mm and cutting edge diameter of 68 mm. Its outside diameters are 72 mm and 74 mm respectively. Determine the area ratio, inside clearance, outside clearance of the sampler. This tube is pushed at the bottom of the bore hole to a distance of 550 mm width with the length of sample being 530 mm. Find the recovery ratio. Comment on the results.

Solution:

Given: $D_1 = 68$mm, $D_2 = 74$mm, $D_3 = 70$mm, $D_4 = 72$mm

Area ratio is given by

$$A_r = \frac{D_2{}^2 - D_1{}^2}{D_1{}^2} \times 100 = \frac{74^2 - 68^2}{68^2} \times 100 = 18.42 < 10\%,\ \text{required for good quality}$$

undisturbed sample.

Inside clearance is given by,

$$C_i = \frac{D_3 - D_1}{D_1} \times 100 = \frac{70 - 68}{68} = 2.94$$

$C_i = 0.5$ to 3% is required for undisturbed sample.

Outside clearance is given by,

$$C_o = \frac{D2 - D4}{D4} \times 100 = \frac{74 - 72}{72} = 2.77$$

$C_o = 0$ to 2% is required for undisturbed sample.

Recovery ratio is given by,

$$R_r = \frac{L}{H} = \frac{530}{550} = 96.4\%$$

$R_r = 96$ to 98% is required for satisfactory undisturbed sample.

Hence, we can say that the given sample is not a good quality undisturbed sample.

Question 15.10: Plot the bore log chart for bore hole at site No. 1 using the data given below:

Project: Shah Nehar Project; Name of structure: Aqua duct at RD 26000.

Depth explored: 7.5 m; Location: Right bank. Position of ground water table: 1.2 m below GL.

Standard penetration resistance values at different depths are:

Depth (m)	0	0.75	1.50	2.25	3.00	3.75	4.50	5.25	6.00	6.75	7.50
Observed N value	0	5	8	10	11	13	15	17	18	22	26

The details of soil tests are:

Depth (m)	IS soil classifi- cation	Description of strata	Grain size analysis			Water content w (%)	Bulk density (g/cm³)	Specific gravity G	$\phi°$
			Sand	Silt	Clay				
1.50	SP	Loose poorly graded sand	95.8	4.2	-	9.2	1.87	2.63	25.8
4.50	SW	Loose to medium well graded sand	96.4	3.6	-	14.2	2.01	2.64	33.6
7.50	SW	Medium to compact well graded sand	96.1	3.9	-	14.2	2.04	2.64	35.4
>7.50		Soft to medium sandstone							

Solution: The bore log chart for bore hole at site No. 1 is plotted as shown in table below.

$$\text{Dry density } \rho_d = \rho/(1 + w) = 1.87/1.092 = 1.71 \text{ g/cm}^3.$$

$$\text{Void ratio } e = (G\rho_w/ \rho_d) - 1 = (2.63 \times 1/ 1.71) - 1 = 0.54$$

Table: Bore hole log chart and data sheet for bore hole at site no. 1.

Project: Shah Nehar Project Position of ground water table: 1.2 m.

Name of structure: Aquaduct at RD 26000   Depth explored: 7.5 m.   Location: Right bank

Depth (m)	Bore log details	IS classification	Observed SPT N value	Observed SPT N-value versus depth plot (0 10 20 30)	Sand (%)	Silt (%)	Clay (%)	Liquid limit (%)	Plastic limit (%)	In-situ water content (%)	Bulk density (g/cm³)	Dry density (g/cm³)	Void Ratio, e	Specific gravity, G	Cohesion, c (kg/cm²)	Φ, degrees	Compression index, C_c	Description of strata
0.0			0															
0.75	GWT	SP	5		95.8	4.2	-	-	-	9.2	1.87	1.71	0.54	2.63	-	25.8	-	Loose poorly graded sand
1.50			8															
2.25			10															
3.00			11					-										
3.75		SW	13		96.4	3.6	-		-	14.2	2.01	1.76	0.50	2.64	-	33.6	-	Loose to medium well graded sand
4.50			15															
5.25			17					-										
6.00		SW	18		96.1	3.9	-		-	14.2	2.04	1.79	0.47	2.64	-	35.4	-	Medium to compact well graded sand
6.75			22															Soft to medium sandstone
7.50			26															

 **Questions**

Question 1: A sampling tube has outer diameter of 80 mm and wall thickness of 1.5 mm. Find the area ratio of the tube and comment on whether the tube could be used for undisturbed sampling of soils.

Question 2: The following sizes of sampling tubes are available in the market:

Sampling tube 1: O.D. = 80 mm, I.D. = 75 mm, length = 550 mm

Sampling tube 2: O.D. = 110 mm, I.D. = 105 mm, length = 550 mm

Sampling tube 3: O.D. = 65 mm, I.D. = 55 mm, length = 550 mm

Out of these which one would you select for obtaining undisturbed soil samples from a borehole if diameter of hole is 100mm.

Question 3: A sampling tube of 150 mm internal diameter is 1 mm thick. It is fitted with the cutting edge. The inside diameter of the cutting edge is flushed with sampling tube. The cutting edge is 1.2 mm thick. Compute inside clearance, outside clearance and area ratio. Comment on sample collected by this tube.

Question 4: The external diameter of a sampling tube is 67 mm. The area ratio required is 11%. Find the thickness of the sampling tube in mm.

Question 5: Compare the area ratio of a thin-walled tube sampler having an external diameter of 60 mm and wall thickness of 2.25 mm. Do you recommend the sampler for obtaining the undisturbed samples? Why?

Question 6: Determine the area ratios of the samplers of the following data and comment on the values.

Diameter	D_o (mm)	D_i (mm)
Split spoon sampler	50	35
Drive tube sampler	100	90
Shelby tube sampler	50	47

Where, D_o = outside diameter of sampling tube, D_i = inside diameter of sampling tube.

CHAPTER 16

Sheet Pile Walls

Question 16.1: Determine the depth of embedment for the cantilever sheet pile shown in the figure 16.1. The soil has an effective unit weight of 17 kN/m³ and an angle of internal friction of 30°. Use a simplified method.

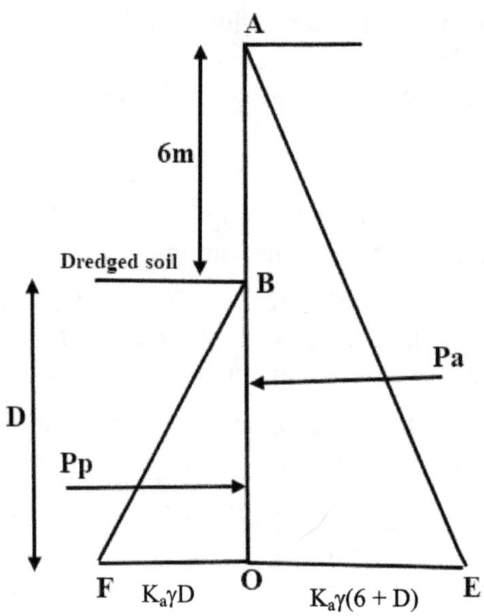

Figure 16.1

Solution:

$$K_a = (1 - \sin\phi)/(1 + \sin\phi) = 1/3$$

$$K_p = (1 + \sin\phi)/(1 - \sin\phi) = 3$$

$$P_a = \text{area of triangle AOE} = \frac{1}{2}\times K_a \times \gamma \times (6+D) \times (6+D)$$

$$P_p = \text{area of triangle BOF} = \frac{1}{2}\times K_p \times \gamma \times D \times D$$

Taking moment about O

$$P_a \times \frac{6+D}{3} - P_p \times \frac{D}{3} = 0$$

$$\frac{1}{2} \times K_a \times \gamma \times (6+D)^2 \times \frac{6+D}{3} - \frac{1}{2} \times K_p \times \gamma \times D^2 \times \frac{D}{3} = 0$$

$$2.833 \times (6+D)^3/3 - 25.5 \times D^3/3 = 0$$

On solving the equation, we get the value of D

D = 5.50m

Actual depth D' = 1.40 × 5.50

$\qquad$ = 7.70m (increase by 40%).

Take D' = 8m (say).

Question16.2: An anchor sheet pile retains dry cohesionless soil to a depth of 5m. The sheet pile is anchored to a depth of 1m from the top. The unit weight of the soil is 18kN/m³ and the angle of internal friction is 30°. Compute the tension in the tie rod per meter length of the pile and the depth of embedment of the pile.

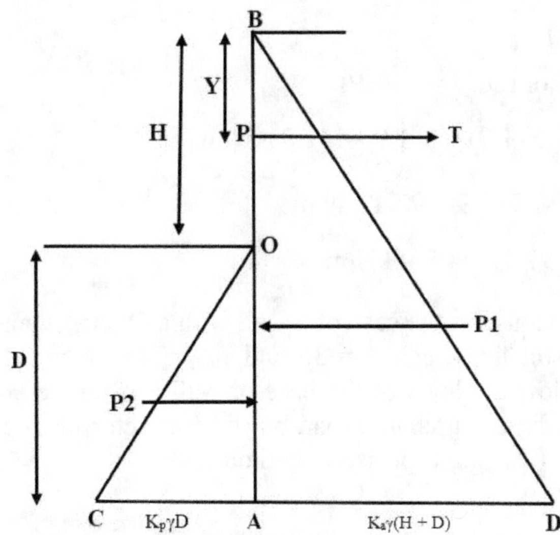

Figure 16.2

Solution:

$$K_a = (1 - \sin 30°)/(1 + \sin 30°) = \frac{1}{3}$$

$$K_p = 1/K_a$$

$$K_p = 3$$

$$Y = 1 \text{ m}$$

From figure 16.2

$$P_1 = \frac{1}{2} \times K_a \times \gamma \times (5+D) \times (5+D)$$

$$P_2 = \frac{1}{2} \times K_p \times \gamma \times D \times D$$

Take moment about the anchor rod level $P = 0$

$$P_1 \times (\frac{2}{3} \times (H+D) - Y) - P_2 \times (2D/3 + (H - Y)) = 0$$

$$P_1 \times (\frac{2}{3} \times (5+D) - 1) - P_2 \times (2D/3 + (5 - 1)) = 0$$

$$\frac{1}{2} \times K_a \gamma (5+D) \times (5+D) \times (\frac{2}{3} \times (5+D) - 1) - \frac{1}{2} \times K_p \gamma \times D \times D \times (2D/3 + (5 - 1)) = 0$$

$$3 \times (5+D)^2 \times (\frac{2}{3} \times (5+D) - 1) - \frac{1}{2} \times 3 \times 18 \times D^2 \times (2D/3 + (5 - 1)) = 0$$

$$3 \times (5+D)^2 \times (\frac{2}{3} \times (5+D) - 1) - 27D^2 \times (2D/3 + (5 - 1)) = 0$$

or D= 1.90m

Taking summation of horizontal force $= 0$;

$$P_1 - P_2 - T = 0$$

Force in anchor rod, $T = P_1 - P_2$

$$P_1 = \frac{1}{2} \times \frac{1}{3} \times 18 \times (5+1.9) \times (5+1.9) = 142.83 \text{kN/m}$$

$$P_2 = \frac{1}{2} \times 3 \times 18 \times 1.9 \times 1.9 = 97.47 \text{kN/m}$$

$$T = 142.83 - 97.47 = 45.36 \text{kN/m}.$$

Question 16.3: Consider the case of a wall with a 2metre retained height of fill constructed in soil for which $\phi = 30°$ and $\gamma_{bulk} = 18$ kN/m^3. The groundwater table is well below the level of the base of wall and ignore any allowance for accidental over dig or surcharge load but take a factor of safety of 2.0 on the passive pressure (Use gross pressure method).

Solution:

Assuming there is no wall friction:

$$K_a = (1 - \sin 30°)/(1 + \sin 30°) = \frac{1}{3}$$

$$K_p = (1 + \sin 30°)/(1 - \sin 30°) = 3$$

Taking moments about point C

Moment due to active force:

$$\text{Moment} = \frac{1}{2} K_a \gamma_{bulk} (h+d)^2 \times (h+d)/3$$

$$= \frac{1}{2} \times \frac{1}{3} \times 18 \times (2 + d)^3 / 3$$

$$= d^3 + 6\,d^2 + 12\,d + 8$$

Moment due to passive force:

$$\text{Moment} = \frac{1}{2}\,K_p\gamma_{bulk}d^2 \times d/3$$

$$= \frac{1}{2} \times 3.0 \times 18 \times d^3/3$$

$$= 9\,d^3$$

Then, equating the active and passive moments and allowing for a factor of safety of 2.0 on the passive gives:

$$4.5\,d^3 = d^3 + 6\,d^2 + 12\,d + 8$$

Which can be solved by trial and error to give d = 3.07 m

The final embedment is then taken as 3.07 × 1.2 = 3.68 metres

It is then required to carry out a force balance to calculate the required value of R for horizontal stability:

$$\text{Active force} = \frac{1}{2}\,K_a\gamma_{bulk}(h+d)^2 = \frac{1}{2} \times \frac{1}{3} \times 18 \times 5.07^2 = 77.11 \text{ kN}$$

$$\text{Passive force} = \frac{1}{2}K_p\gamma_{bulk}\,d^2/2 = \frac{1}{2} \times 3.0 \times 18 \times 3.07^2/2 = 127.24 \text{ kN}$$

Hence, R = 127.24 – 77.11 = 50.13 kN

This force must be generated in the additional length of pile below point C.

The two force elements generated will be due to the passive load on the back of the wall and the active force on the front of the wall (see figure 1(b), above). In each case there will be two elements – one due to the effective surcharge imposed by the fill above the level of point C and one due to the soil self-weight below level C.

Overall, the lateral effective stress profile will be as shown below:

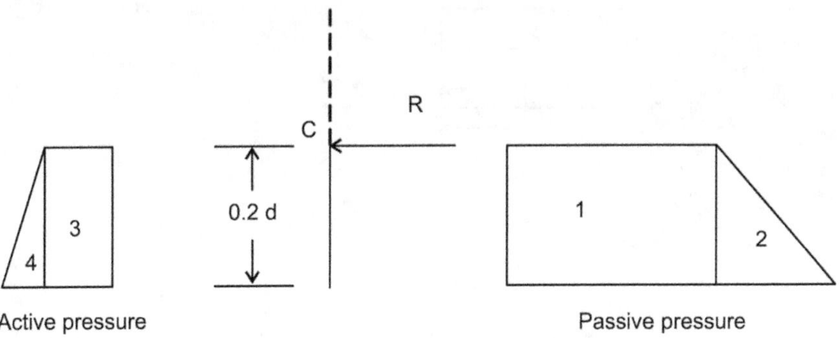

Figure: 16.3

Refer to figure 16.3

Passive pressures:

Force 1, due to overburden surcharge = $K_p \times \gamma_{bulk} \times (h+d) \times 0.2\,d$

$$= 3 \times 18 \times 5.07 \times 0.2 \times 3.07 = 168.1\ kN$$

Force 2, due to soil self-weight = $0.5 \times K_p \times \gamma_{bulk} \times (0.2\,d)^2$

$$= 0.5 \times 3 \times 18 \times (0.2 \times 3.07)^2 = 10.18\ kN$$

Active pressures:

Force 3, due to overburden surcharge = $K_a \times \gamma_{bulk} \times d \times 0.2\,d$

$$= \frac{1}{3} \times 18 \times 3.07 \times 0.2 \times 3.07 = 11.31\ kN$$

Force 4, due to soil self-weight = $0.5 \times K_a \times \gamma_{bulk} \times (0.2\,d)^2$

$$= 0.5 \times \frac{1}{3} \times 18 \times (0.2 \times 3.07)^2 = 1.13\ kN$$

Available force = passive − active = $168.10 + 10.18 - 11.31 - 1.13 = 165.84$ kN

Required force = 50.13 kN < available force of 165.84; hence, OK.

Question 16.4: Determine the depth of embedment for the cantilever sheet pile shown in figure. The soil has effective unit weight of 17 kN/m^3 and an unconfined compressive strength of 70 kN/m^2. Use simplified method.

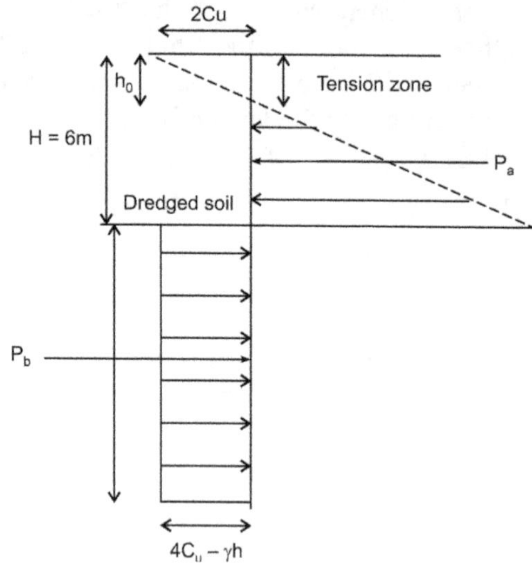

Figure 16.4

Solution:

$q_u = 70\ kN/m^2$

$$C_u = \frac{q_u}{2} = 35\ kN/m^2$$

$$h_o = \frac{2Cu}{2} = \frac{2 \times 35}{17} = 4.12 \text{ m}$$

$$K_a = \frac{1 - sin\phi}{1 + sin\phi} = 1 = K_p$$

$$P_a = \frac{1}{2}[K_a\gamma\, H - 2C_u\sqrt{K_a}\,] \times [H - h_0]$$

$$P_a = \frac{1}{2}[1 \times 17 \times 6 - 2 \times 35 \times 1] \times [6 - 4.12]$$

$$P_a = 30.08 \text{ kN/m}$$

$$P_p = (4C_u - \gamma\, H)\, D$$

$$P_p = (4 \times 35 - 17 \times 6) \times D$$

$$P_p = 38D$$

$$\Sigma M_D = 0$$

$$P_a \times [D + [\tfrac{H - ho}{3}]] = P_p \times\frac{D}{2}$$

$$30.08 \times [D + 0.6267] = \frac{38}{2}\times D^2$$

$$D = 2.063 \text{m}$$

Actual depth $D' = 1.4 \times D = 2.88$m

or take $D' = 3$m.

Question 16.5: Determine the depth of embedment of the given anchored sheet pile by free earth support method. Also determine the force in the anchor per meter of the wall. Assume the following soil parameters: $\gamma = 16 \text{ kN/m}^3$, $\gamma_{sub} = 9 \text{ kN/m}^3$, $\phi = 35°$.

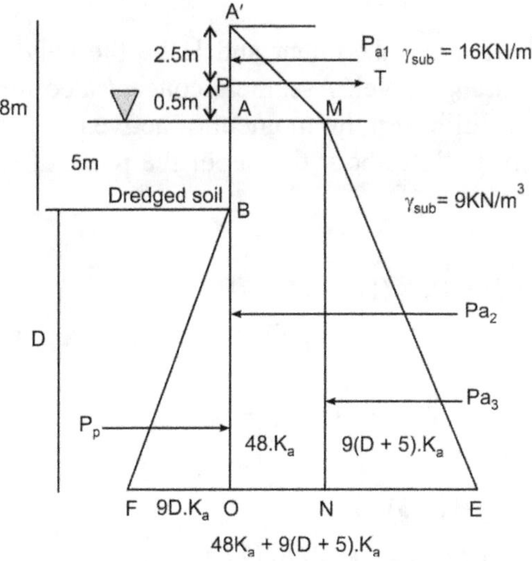

Figure 16.5

Solution:

$$k_a = \frac{1-\sin\phi}{1+\sin\phi} = 0.27 \qquad k_p = \frac{1+\sin\phi}{1-\sin\phi} = 3.69$$

$$\Sigma M_p = 0$$

$$P_{a1} \times (2.50 - \tfrac{2}{3} \times 3) - P_{a2} \times (\tfrac{D+5}{2} + 0.50) - P_{a3} \times [(\tfrac{2}{3} \times (D+5) + 0.50)] + P_p \times (\tfrac{2}{3} \times D + 5.50) = 0$$

$$10.256D^3 + 72.014D^2 - 132.64D - 312.03 = 0$$

$$D = 2.57\text{m}.$$

$$P_{a1} + P_{a2} + P_{a3} - P_p - T = 0$$

$$T = P_{a1} + P_{a2} + P_{a3} - P_p$$

$$T = 19.512 + 13.008 \times (2.57+5) + 1.2195 \times (2.57+5)^2 - 16.605 \times 2.57^2$$

$$T = 78.19\text{kN/m}$$

$$P_{a1} = \tfrac{1}{2} \times 48\, K_a \times 3 \times 3 = 19.512 \text{ kN/m}$$

$$P_{a2} = 48\, K_a \times (D+5) = 13.008\, (D+5) \text{ kN/m}$$

$$P_{a3} = \tfrac{1}{2} \times K_a \times 9 \times (D+5) \times (D+5) = 1.2195 \times (D+5)^2 \text{kN/m}$$

$$P_p = \tfrac{1}{2} \times K_p \times 9 \times D \times D = 16.605D^2 \text{kN/m}$$

$$D' = 1.4 \times 2.57 = 3.60\text{m}$$

Question 16.6: With suitable illustrations describe the simplified analysis method for designing the depth of embedment of the cantilever sheet pile for a 6m deep excavation in a sandy soil layer for $\gamma = 18\text{kN/m}^3$ and $\phi = 35°$ for a factor of safety of 2.

Solution:

Let D be the depth of the embedment and H be the height of the cantilever sheet pile above the dredge level. Assume a concentrated force R acting at the foot of the pile. For equilibrium the moment of active pressure on the right and passive resistance on the left about the about the piont of reaction R must be balanced.

$$\Sigma M = 0$$

$$P_p \times (D/3) - P_a \times (H + D/3) = 0 \qquad\qquad (1)$$

We will provide F.O.S = 2, against passive force which is providing stability.

$$(P_p/2)(D/3) - P_a \times (H + D)/3 = 0$$

$$P_P = \tfrac{1}{2}\, K_p \gamma D^2$$

$$Pa = \tfrac{1}{2}\, K_a \gamma (H+D)^2$$

Substituting the value of P_p and P_a in (1)

$$(1/6) \times (\tfrac{1}{2})\, K_p \gamma D^2 \times D - (H + D)/3 \times \tfrac{1}{2}\, K_a \gamma (H+D)^2 = 0$$

$$\gamma/6 \times [K_pD^3/2 - K_a(H+D)^3] = 0$$

$$K_pD^3 - 2K_a(H+D)^3 = 0$$

Data given

$\phi = 35°$ $H = 6m$

$\gamma = 18kN/m^3$

$F.O.S = 2.0$

$$K_a = \frac{1-sin\phi}{1+sin\phi} = \frac{1-sin35°}{1+sin35°} = 0.271$$

$$K_p = 1/K_a = 3.69$$

$$3.69D^3 - 2 \times 0.271 \times (6+D)^3 = 0$$

$$3.69D^3 - 0.542 \times [(6)^3 + D^3 + 3 \times 6D(D+6)] = 0$$

$$3.148D^3 - 9.756D^2 - 58.536D - 117.072 = 0$$

By trial and error, $D = 6.71m$

Therefore, depth of embedment for sheet pile $D = 6.71m$.

Questions

Question 1: Acantilever sheet pile retains soil to a height of 5m. Find the depth to which the pile should be driven assuming two-third of the theoretical passive resistance is developed on the embedded length. $\gamma = 21$ kN/m^3 and $\phi = 25°$. Use approximate method.

Question 2: Determine the depth of embedment for the cantilever pile in clay, shown in given figure 16.6. The water table is at a height of 2.5m above the dredged level on both sides.

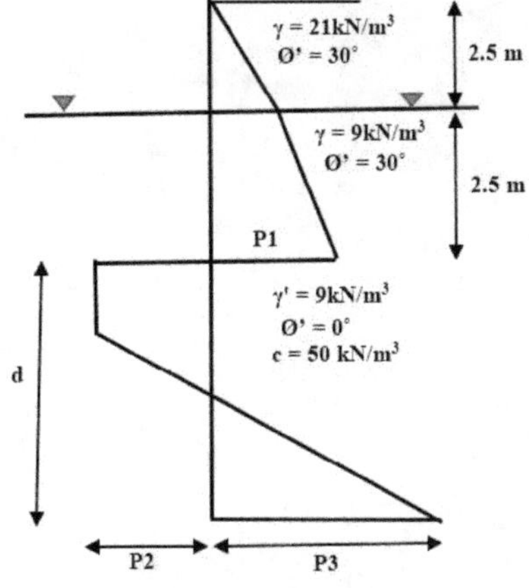

Figure 16.6

Question 3: For the anchored sheet pile shown the figure 16.7 given below, determine the theoretical and actual depth of penetration, the anchor force per unit length of the wall and maximum moment. Assume free-earth support.

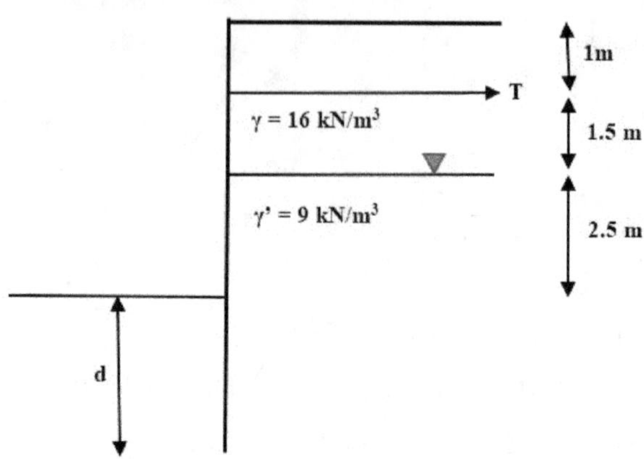

Figure 16.7

Question 4: Consider the case of a wall with a 3m retained height of fill constructed in soil for which $\phi = 25°$ and $\gamma_{bulk} = 21$ kN/m^3. The groundwater table is well below the level of the base of the wall and ignoring any allowance for accidental over dig or surcharge load but takes a factor of safety of 1.5 on the passive pressure (Gross pressure method).

Question 5: An excavation of 7m depth is to be made in cohesionless soil ($\gamma = 21$ kN/m^3, $\phi = 25°$). The sides of the excavation are supported by anchored sheet piles with fixed-end support. Determine the minimum depth of embedment for equilibrium. The anchors are at a depth of 2 m below the top surface.

References

1. Bureau of Indian Standards (1973) Methods of Tests for Soil, Part 2, "Determination of Water Content of Soil". IS 2720, B.I.S, New Delhi.

2. Bureau of Indian Standards (1985) Methods of Tests for Soil, Part 4, "Grain Size Analysis". IS 2720, B.I.S, New Delhi.

3. Bureau of Indian Standards (1985) Methods of Tests for Soil, Part 5, "Determination of Liquid Limit and Plastic Limit". IS 2720, B.I.S, New Delhi.

4. Bureau of Indian Standards (1980) Methods of Tests for Soil, Part 3/Sec1: "Determination of Specific Gravity". IS 2720, B.I.S, New Delhi.

5. Bureau of Indian Standards (1992) Methods of Tests for Soil, Part 7, "Determination of Water Content-Dry Density Relation using Light Compaction of Soil". IS 2720, B.I.S, New Delhi.

6. Bureau of Indian Standards (1991) Methods of Tests for Soil, Part 10, "Determination of unconfined compressive strength of soil". IS 2720, B.I.S, New Delhi.

7. Bureau of Indian Standards (1986) Methods of Test for Soil, Part 17, "Laboratory determination of permeability". IS 2720, B.I.S, New Delhi.

8. Bureau of Indian Standards (1975) Methods of Tests for Soil, Part 29, "Determination of dry density of soils in-place by the core cutter method". IS 2720, B.I.S, New Delhi.

9. Bureau of Indian Standards (1986) Methods of Tests for Soil, Part 15, "Determination of consolidation properties". IS 2720, B.I.S, New Delhi.

10. Bureau of Indian Standards (1980) Methods of Test for Soils, Part XXX, "Laboratory vane shear test". IS 2720, B.I.S, New Delhi.

11. Bureau of Indian Standards (1986) Methods of Tests for Soil, Part 13, "Direct shear test". IS 2720, B.I.S, New Delhi.

12. Bureau of Indian Standards (1993) Methods of Tests for Soil, Part 11, "Determination of the shear strength parameters of a specimen tested in unconsolidated undrained triaxial compression without the measurement of pore water pressure". IS 2720, B.I.S, New Delhi.

13. Bureau of Indian Standards (1981) Methods of Tests for Soil, Part 12, "Determination of the shear strength parameters of soil from consolidated undrained triaxial compression test with measurement of pore water pressure". IS 2720, B.I.S, New Delhi.

14. Bureau of Indian Standards (1981) Methods of Tests for Soil, Part 12, "Determination of the shear strength parameters of soil from consolidated undrained triaxial compression test with measurement of pore water pressure". IS 2720, B.I.S, New Delhi.

15. Bureau of Indian Standards (1970), Methods of Tests for Soil, Edition 2.2, "Soil Classification System", IS 1498, B.I.S, New Delhi.

16. Bureau of Indian Standards (1977) Methods of Test for Soil, Part XL, "Determination of Free Swell Index of Soils". IS 2720, B.I.S, New Delhi.

17. Bureau of Indian Standards (1987) Methods of Test for Soil, Part 26, "Determination of pH value". IS 2720, B.I.S, New Delhi.

18. Bureau of Indian Standards (1972) Methods of Test for Soil, Part VI, "Determination of Shrinkage factors". IS 2720, B.I.S, New Delhi.

19. Bureau of Indian Standards (1981) (Reaffirmed 2002), "Method for Standard Penetration Test for Soils", IS 2131, B.I.S, New Delhi.

20. Bureau of Indian Standards (1982) (Reaffirmed 2002), "Method of Load Test on Soil", IS 1888, B.I.S, New Delhi.

21. Bureau of Indian Standards (1976) (Reaffirmed 2007), "Method for Subsurface Sounding for Soils", IS 4968 (Part I), B.I.S, New Delhi.

22. Bureau of Indian Standards (1976) (Reaffirmed 2007), "Method for Subsurface Sounding for Soils", IS 4968 (Part II), B.I.S, New Delhi.